SOJOURNING IN DISCIPLINARY CULTURES

SOJOURNING IN DISCIPLINARY CULTURES

A Case Study of Teaching Writing in Engineering

EDITED BY
MAUREEN A. MATHISON

UTAH STATE UNIVERSITY PRESS
Logan

© 2019 by University Press of Colorado

Published by Utah State University Press
An imprint of University Press of Colorado
5589 Arapahoe Avenue, Suite 206C
Boulder, Colorado 80303

All rights reserved

 The University Press of Colorado is a proud member of the Association of University Presses.

The University Press of Colorado is a cooperative publishing enterprise supported, in part, by Adams State University, Colorado State University, Fort Lewis College, Metropolitan State University of Denver, Regis University, University of Colorado, University of Northern Colorado, Utah State University, and Western State Colorado University.

ISBN: 978-1-60732-802-5 (paperback)
ISBN: 978-1-60732-803-2 (ebook)
https://doi.org/10.7330/9781607328032

Library of Congress Cataloging-in-Publication Data

Names: Mathison, Maureen A., editor.
Title: Sojourning in disciplinary cultures : a case study of teaching writing in engineering / edited by Maureen A. Mathison.
Description: Logan : Utah State University Press, [2018] | Includes bibliographical references and index.
Identifiers: LCCN 2018041473 | ISBN 9781607328025 (pbk.) | ISBN 9781607328032 (ebook)
Subjects: LCSH: English language—Rhetoric—Study and teaching (Higher) | Interdisciplinary approach in education. | Technical writing—Study and teaching (Higher) | Academic writing—Study and teaching (Higher) | Communication in engineering.
Classification: LCC PE1404.S633 2018 | DDC 808.06/6621—dc23
LC record available at https://lccn.loc.gov/2018041473

The University Press of Colorado gratefully acknowledges the generous support of the College of Humanities and the Department of Writing and Rhetoric Studies at the University of Utah toward the publication of this book.

Cover illustration © Mertsaloff/Shutterstock.com

*To my daughter, Marianna,
who, at the tender of age of two,
asked if the clouds were the moon's jewelry.*

CONTENTS

Acknowledgments ix

Introduction: Integrating Writing throughout One College, Many Departments
Maureen A. Mathison 3

1. Sojourners and Third Cultures: Raising Cultural Awareness in Interdisciplinary Programs
Maureen A. Mathison and Mara K. Berkland 13

2. Professors Designing Assignments as Relational Activity: A Baseline for Connecting Thinking, Learning, and Writing
Maureen A. Mathison and Linn K. Bekins 36

3. Teaching (Each Other) (about) Writing
Doug Downs 50

4. Locating Common Ground for Diplomacy: Using Critical Thinking to Teach Writing
Sarah Read and Maureen A. Mathison 71

5. Moving toward Successful Interdisciplinary Integration in Team-Taught Courses: Building Cultural Bridges through Assignments
Mara K. Berkland 87

6. "I Don't Have to Argue My Design—The Visual Speaks for Itself": A Case Study of Mediated Activity in an Introductory Mechanical Engineering Course
Maureen A. Mathison 105

7. I See What You Mean: Mechanical Engineering Students' Use of Visuals in a Research Paper Assignment
Sarah A. Bell 118

8   Ideologies of Gender: Culture Clash between the Disciplines
    April A. Kedrowicz and Julie L. Taylor   134

9   Intercultural Collaboration: Respect, Relationship, Responsibility, and Reciprocity
    Sundy Watanabe   154

10 Sojourning, Resistance, and Trust
    Maureen A. Mathison   175

   References   189
   About the Authors   207
   Author Index   209
   Subject Index   211

ACKNOWLEDGMENTS

Writing is one of the most powerful means of enacting change. And yet, it is through our interactions with people that our writing can achieve its potential. Many people are owed gratitude for helping me with this project at various points in time and through its various stages. First, there would have been no project had Bob Roemer, then chair of the Department of Mechanical Engineering, not telephoned (that's how long ago the seeds for this program were sown!) Ann Darling, a colleague in the Communication Department, and myself to assist with an Accreditation Board for Engineering and Technology (ABET) visit. Together the three of us met regularly to brainstorm and plan an exciting educational reform in integrating writing and speaking throughout the curriculum in the department. I learned much from them about the sharing of ideas and their implementation into practice. The program would not have succeeded without the intelligence, commitment, and efforts of so many graduate students who willingly traveled to another disciplinary culture. In the very early years, Deanna Dannels and Stephanie Mackay were instrumental in providing their expertise. Later, the authors in this volume carried the project forward as it expanded into other departments in the College of Engineering. Of special note is the leadership of April Kedrowicz, who eventually became the director of the program and showed great insight and strength in her ability to collaborate with so many faculty members from the College of Engineering, and to administer one of the most unique and extensive writing in the disciplines/communication in the disciplines (WID/CID) programs in the country.

I believe a project such as this, at the time it was conceived in the 1990s, would only have been possible at an institution such as the University of Utah, where administration and faculty have generally been open to innovation and change. I am still convinced that the decision I made twenty-odd years ago to uproot and move west of the Mississippi was a stroke of good fortune. Over the years, I have been surrounded by the best colleagues: John Ackerman, Jenny Andrus, Casey Boyle, José Cortez, Dan Emery, Romeo García, Tom Huckin, Jay Jordan, LuMing Mao, Susan Miller (deceased), Joy Pierce, Raúl Sánchez, Jon Stone, and

Christie Toth. Some have moved on, but all have sustained me with their conversations and support and have always made me eager to go into the office.

The volume itself was initially formulated at my kitchen table when Doug Downs, Sarah Read, and I were discussing our learning experiences as embedded consultants. It solidified at a lunch with Susan Miller, who suggested it was timely. Many of the chapters were presented at various conferences (College Composition and Communication, Rhetoric Society of America, Western States Rhetoric and Literacy Conference, and Writing Research across Borders) and have benefited from feedback from attending audiences. A recent sabbatical in the College of Humanities at the University of Utah afforded me the time to direct my complete attention to the project. I am grateful to the reviewers of the volume, who offered keen readings and made insightful and constructive comments that ultimately brought it to fruition. Michael Spooner, acquisitions editor, saw the project in its infancy and had great patience as he guided me through the process with encouragement. To him and the staff at the press, thank you.

I also thank Judith Mathison and Michael Mathison, siblings who have entertained my musings regarding the volume and encouraged me along the way. Finally, my daughter, Marianna, is owed a huge "thank you" for the luster and texture she adds to my life.

SOJOURNING IN DISCIPLINARY CULTURES

Introduction

INTEGRATING WRITING THROUGHOUT ONE COLLEGE, MANY DEPARTMENTS

Maureen A. Mathison

This volume examines Writing in the Disciplines (WID) from a cultural theoretical perspective (Becher and Trowler 2001; Klein 2009), reporting on a collaboration between writing and engineering to develop a model undergraduate program in which writing was integrated throughout the curriculum. To date research has tended to either (1) emphasize the challenges students face when writing in their discipline or (2) emphasize the challenges WID instructors face when collaborating with those outside their discipline. We focus on the second, less-examined challenge, proposing that tension is a normal aspect of collaborating between disciplines. Specifically, the chapters in this volume address dissonant areas of cultural assumptions and dispositions between the "hard" and "soft" disciplines of engineering and writing and how they were negotiated or ameliorated.

In *Learning to Communicate in Science and Engineering: Case Studies from MIT*, Mya Poe, Neil Lerner, and Jennifer Craig (2010) identify and elaborate some of the challenges engineering students encounter as they learn to write for an audience of peer engineers. While the book focuses on student learning, the authors also claim that they, too, learned through their experience, though they provide little elaboration. "At times," they remark, the collaboration between writing and engineering was "frustrating, as our values and background knowledge seemed so disparate" (199). Likewise, Lea Anna Cardwell (2016), managing editor of the *WAC Journal*, similarly points out that WAC/WID work can be tenuous, as demonstrated when she wrote a call for papers for a volume on "concerns or problems" in writing across the curriculum (WAC). This collection speaks to both Poe's et al. and Cardwell's comments, illuminating how interdisciplinary collaboration is a coming together of different values and perspectives; the timely chapters in this volume examine

DOI: 10.7330/9781607328032.c000

some of the areas where particular misunderstandings occurred during collaboration between members of a College of Humanities and a College of Engineering at a large public research institution.

The volume represents the collective experiences and insights of writing consultants involved in a large-scale curriculum reform of an entire college of engineering; they collaborated closely with faculty members of the various departments and taught writing in engineering classrooms to engineering students. The unique project was initiated in 1996, when, in anticipation of an accreditation review, the chair of the Department of Mechanical Engineering contacted me to request help in improving his students' writing. After several meetings with him, I decided that since I was unfamiliar with the knowledge of the discipline and the specific culture of the department (see Godfrey and Parker 2010; Pawley 2009 for descriptions of engineering culture and beliefs), it would be best to move slowly. There would be no quick fixes; instead, I spent a year as an ethnographic observer attending undergraduate design sequence classes in the Department of Mechanical Engineering and taking notes as if I were an engineering student. Through immersing myself in the classroom talk and conversations, I came to have a better, yet imperfect, understanding of how writing could be incorporated into a department-wide curriculum, not as an "add-on," but as an integral component of courses. After one year, I had a sense of where writing might be most useful in the design curriculum and at what points for specific learning goals. Six courses were selected, from the introductory first-year design course through the capstone.

The goal was to simulate—as much as is possible—the demands of workplace writing in these courses (see Dannels 2000). Employing a situated learning model, writing became part of the practice of engineering for students. Whereas before they had writing assignments scattered throughout the curriculum, the new curriculum implemented writing at strategic engineering teaching moments, starting with easier genres and graduating to more difficult ones, spiraling throughout the curriculum. With this scaffolding, zones of proximal development were created to support students as they progressed through the curriculum and increased their knowledge of practices (Vygotsky 1978); writing assignments in one course prepared them to undertake more difficult writing assignments in future courses. For example, in an early design course, students learned to write memos to their "manager" as a means of keeping him or her informed about their progress on a project. A course taught after the design course still included memos but added a feasibility report in which students responded to a request for proposals

(RFP) to build, say, a grease trap. Writing assignments and engineering content were concomitant, and as students progressed through the curriculum, both became more complex, while reinforcing previously learned genres. Earlier work set the stage for later work so that students could learn new genres as they encountered new contexts of practice while strengthening previously learned genres. Genres are powerful instructional tools; as Marie Paretti (2008) notes, "analyses of genres in academic, government, and industry sites have provided compelling insights into the ways in which the structure, tone, content, organization, and related features of documents support the human activities to which those documents respond" (493). Having students write throughout their course work supported their becoming more prepared for activities in the workplace.

My own observations and suggestions were mediated through regular meetings with engineering faculty in the department, who in collaboration with me created the curriculum for the design sequence in mechanical engineering so that students would strategically learn about and practice specific genres of engineering communication that were relevant to their professional identities. Over the course of two years, I collaborated closely with the mechanical engineering professors whose courses were included in the new curriculum. We developed syllabi that incorporated writing into their courses in meaningful ways, designed lessons to teach various aspects of writing, created assignments that integrated engineering and writing theory and concepts, and worked one on one with students to provide feedback for revision. In the first year of implementation, I was the sole writing consultant embedded within engineering. I attended classes and provided minilessons about the type of writing students were being asked to complete and how it was relevant to the workplace. Minilessons were critical, as the engineering professor who was in the classroom could elaborate on engineering practices and provide stories about workplace life and the positioning of writing within it. I also attended labs and met with student teams to support their learning of the relevant genres and to provide feedback. The initial year was intensive, with me spending approximately six to eight hours weekly in mechanical engineering while maintaining my normal load in my own department.

As the relationship between writing, communication (there was also an oral component of the program), and engineering developed, we scrambled to fund graduate students who had a keen interest in the technological sciences and writing and who wanted to fulfill their graduate teaching assistantship full-time in the department. For many years we drew from whatever resources on campus we could to fund graduate writing consultants.

With ABET 2000, or EC 2000 as it is also known, the imperative for a wider program integrating communication across the college of engineering curriculum became more pronounced. The newer outcome criteria for undergraduate engineering education called for students to demonstrate "the ability to communicate effectively." These criteria, explains Carolyn Miller (2004), were not ranked in terms of importance, highlighting the potential synergies among them. Furthermore, "the criteria emphasized that communication is a strategic, situated enterprise that must be judged in context and with an understanding of the constraints and conventions in play and of the challenges to be met" (42). The goals of the program we had begun years before aligned with the new criteria and were highly valued in the workplace.

In light of EC2000, and with a successful curriculum in place in mechanical engineering, we became one of nine engineering colleges under the Engineering Schools of the West Initiative (ESWI). As part of this group, in 2003, we were awarded a $1.1 million William and Flora Hewlett Foundation Grant over five years that allowed us to expand and solidify the established program in mechanical engineering to house a "Center" across four additional departments: bio; chemical; civil and environmental; and materials science. In addition to writing, oral communication and ethics were also part of the engineering curriculum. Other institutions in ESWI received varied funding for projects that ranged from hosting summer camps for middle and high school students, to working with science teachers (for a description of the larger project, see Plumb and Reis 2007).

Our program was unique because the teaching and learning of writing and oral communication occurred *in* engineering classrooms and not as separate courses. Our program was similar to a writing fellow program but differed in significant ways. In their introduction to a special volume in *Across the Disciplines*, Brad Hughes and Emily Hall (2008) explain that while such programs vary in their implementation of goals, they commonly "link students to specific writing-intensive courses, they encourage partnerships between a [w]riting [f]ellow and a course professor; and they promote collaboration between peers" (1). The origin of fellow programs traces back to Tori Haring–Smith, who after first establishing a drop-in writing center developed the first writing fellow program in the country in 1982 at Brown University. According to Haring-Smith (1992), the goal of the program was to provide support to undergraduate students for courses across the curriculum. Using undergraduate students trained to provide peer support, fellows worked with faculty who represented a range of academic disciplines to better understand their assignments, particularly their purpose and expectations for performance. The information garnered from these meetings helped fellows provide more targeted feedback to students on their drafts and in conferences. Haring-Smith made it clear that writing fellow programs differ from writing centers in that rather than students coming to one central location, fellows go into classrooms and work with faculty and their students. Throughout the duration of a term, faculty and fellows communicate to apprise each other of key information or needs to sustain relationships and enhance writing quality. Other institutions have since established their own writing fellow programs, adapting them to their unique educational contexts with common outcomes (McLeod and Soven 2000; Mullen 2008; Thaiss and Zawacki 2006), including deeper appreciation of, and increased attention to, writing from faculty across the disciplines and improving faculty-writing relationships. Of great importance, research shows that writing fellow programs can have a positive impact, improving the quality of student writing (Rossman-Regaignon and Bromley 2011), including that of nonnative speakers of English (Manley 2014).

Our program distinguished itself from writing fellow programs because it (1) was intended to reform the curriculum of a college and not just one course; (2) was focused on one field of study, engineering, incorporating its subdisciplines; (3) was established using a situated learning perspective, where students learn best in the situations for which the information and practices are relevant; (4) assigned graduate student consultants to an entire department, rather than undergraduate peer fellows to a single course; (5) assigned graduate students to the same department and courses over two or more years; and (6) housed the consultants in the College of Engineering; their offices were located in the building that housed other engineering faculty. They were also paid through engineering. This configuration of characteristics made them "Sojourners," travelers to a new disciplinary culture (explained in more depth in chapter 1). Conceptually the program shared many of the same attributes of a writing fellow program, but had more in common with the one at CUNY that Mary Soliday (2011) describes in *Everyday Genres: Writing Assignments across the Disciplines*. Like Soliday, our guiding theory was that of situated cognition, viewing engineering students at our institution as apprentices; and our focus was also on the teaching and learning of genres. CUNY embedded graduate students in classrooms to teach and collaborate alongside professors, as did we. Both programs facilitated change in assignments and the support of students in fulfilling them successfully through conferencing. The programs also had marked differences.

Whereas the graduate fellows at CUNY represented various fields (e.g., education, music), our graduate students were either advanced MA or PhD students in rhetoric and writing studies and had a level of expertise in writing theory and practice. Unlike CUNY, where graduate fellows were selected to work with one faculty member, our goal was to impact an entire academic college rather than a single course or faculty member. To do this meant collaborating with faculty across the College of Engineering to establish a novel, reciprocal model whereby graduate consultants in writing learned about engineering as faculty in engineering learned about writing. The two expertises were exploited to develop new territory. That is, our fellows, called "consultants" because of the expertise they brought to the project, did not go in and solely work in one class with extant assignments to improve them, or create new assignments as Soliday's did, but collaborated with multiple faculty and their courses in one entire department to develop curricula, teach, and support writing at strategic learning moments for students. In many cases, writing was integrated into key courses that did not previously include it to create a coherent and more seamless curriculum.

It should be noted that every department was treated as its own unique culture given its purpose, history, and practices are distinct from each other. Assignments developed for mechanical engineers would not likely transfer to bioengineers, who operate with different scientific theories, applications, and goals. Audiences vary across engineering subdisciplines, as do their rhetorical means of persuasion. They have their own professional associations (American Institute of Chemical Engineers vs. American Association for Engineers), journals (*Journal of Material Science* vs. *Biotechnology and Bioengineering*); conferences (American Society of Mechanical Engineers Conference vs. Electrical Transmission and Substation Structures Conference). and use of specialized language and visuals. In effect, each subdiscipline of engineering represents a separate discourse community (see Swales 1990). This can become complicated because subdisciplines are even more fine-grained when considering specialties. This is made clear in Thaiss and Zawacki's (2006) book *Engaged Writing, Dynamic Disciplines: Research on the Academic Writing Life*, when they interview professors about their areas of expertise: "Regan, although naming her discipline 'political science,' said she could identify 40 distinct branches of the field, each with its own journals and standards, and saw her own work as 'technology studies,' distinct, say, from 'policy studies'" (34).

The authors in this volume were some of the very first writing graduate students to be placed into the different departments of engineering; they worked closely with faculty to design a curriculum that rang true to the principles of writing and rhetoric and to the specific cultures of the departments of engineering into which they were placed. No one had the same experience, though some had similar frustrations. They were "on the ground," so to speak, at the inception of the college-wide program. Being present at the beginning allowed them to experience firsthand the tensions that arose during the implementation of organizational change (Faber 2002). More often than not, contrasting beliefs about writing and its role in engineering came to the forefront when collaborating. What were the beliefs? How were they engaged? How were they negotiated?

While programs that integrate writing into engineering curricula are unique, the experience of entering a new academic culture so different from one's own is not. Accounts of tensions and incongruities across disciplines abound in the literature. With the exception of Chris Anson (2002), however, few are based in on-the-ground experience. His casebook is helpful in creating awareness of the tensions and posing questions about how to respond in such situations. The current volume, like Anson's, acknowledges that tensions arise. They are a fact of interdisciplinary collaborations. While they may be uncomfortable to engage, they are healthy in that they signal a pathway to improve collaboration and its goals. Tensions indicate differences in epistemologies and ideologies and their constitutive practices. They serve as references to different histories and trajectories of disciplines. And they provide critical points for understanding.

In their study about successful collaborations across disciplines, Maura Borrego, Lynita Newswander, and Lisa McNair (2007) comment that the ability to be open to and appreciate different views about knowledge is important, especially when the disciplines collaborating hold very different beliefs about it. But it may take time for the appreciation to develop. When instructors of writing collaborate across disciplines, unequivocally they will, as Michelle Fine says, hit "speed bumps," at least in the early stages of collaboration. Speed bumps, she explains are "raised places in the road that limit one's speed. When we are moving too fast, we must suddenly slow down or be thrown off course" (1). And so it was with the writing consultants, sojourners traveling to another discipline, confident in their own cultural beliefs and values, but unfamiliar with those in engineering. Change was slow in integrating writing and engineering; moving too quickly would have thrown us off course (and still, moving slowly, we sometimes were).

This volume is relevant for those interested in pursuing WAC/WID (potentially CID), either as newcomers or in the early stages of collaboration, and for those interested in implementing an extensive program like ours. The book addresses interdisciplinary teaching from various perspectives, with each chapter taking up an issue related to collaborating between disparate disciplines. Through a variety of styles and methods, the volume relays the first years of the program. The first two chapters furnish the background for the project. Chapter 1, "Sojourners and Third Cultures: Raising Cultural Awareness in Interdisciplinary Programs," maps out the theoretical foundation of the volume and project. Employing intercultural communication theory, Maureen A. Mathison and Mara K. Berkland theoretically examine disciplines as cultures and address five issues that can impede successful interdisciplinary collaboration. Chapter 2, "Professors Designing Assignments as Relational Activity: A Baseline for Connecting Thinking, Learning, and Writing," applies an activity theory perspective, illuminating how writing was initially situated in the College of Engineering before the collaborative project commenced. Through interviews with professors and analysis of their course materials, Maureen A. Mathison and Linn K. Bekins found distinct approaches to activities in their classrooms, with some more aligned with writing theories and practices than others. The interviews also provided a rough baseline in anticipation of the ways writing was being addressed in distinct departments when consultants entered into their classrooms.

The next three chapters examine how graduate student writing consultants engaged engineering faculty as curricula were revised. This was a major endeavor, particularly at the beginning of our relationship, when our differences became visibly and viscerally apparent. The chapters, in the words of Melinda Whitfield (2014), are told "through the voices of the story-telling authors," a rich narrative style (239) that recounts their "real-life experiences" (238). This is apparent as each author narrates his or her experience. The chapters comprise different richly textured voices, each author accounting for their distinctive collaborative style. They write of concerns relative to their position, and employ varied theoretical lenses and methodologies for their analyses. Combined they provide insight into the multilayered complexity of collaboration between WAC/WID programs and their academic partners. In chapter 3, "Teaching (Each Other) (about) Writing," Doug Downs focuses on how cultural conceptions of writing between those in writing (rhetoric) and civil and environmental engineering (scribal) differ and how those conceptions served as the basis for teaching each other about disciplinary assumptions. Sarah Read and Maureen A. Mathison, in "Locating Common Ground for Diplomacy: Using Critical Thinking to Teach Writing" (chapter 4), recount the role of establishing diplomacy as a writing consultant, and the Chemical Engineering faculty collaboratively designed a curriculum that ultimately became a guiding document for writing throughout the department. The next chapter, "Moving Toward Successful Interdisciplinary Integration in Team-Taught Courses: Building Cultural Bridges through Assignments" (chapter 5), by Mara K. Berkland, demonstrates how the theoretical and pedagogical interests of those in writing and mechanical engineering came together to create assignments that resonate for both engineers and writing consultants. These chapters demonstrate, to differing degrees, the levels of miscommunication experienced as cultures collided, and as collaborators stretched to understand the other.

The next series of chapters are research based, drawing on qualitative and descriptive methods. Chapters 6 and 7 address a key hallmark of engineering practice: visual approaches to conceptual understanding. Through their training, writing instructors have been more immersed in the verbal aspects of teaching and learning (though this is rapidly changing), whereas engineering often relies on the visual. To connect the verbal and the visual, writing instructors used the visual as a starting point. How do engineers view the role of visuals in writing, and how do students integrate them into their engineering work? In chapter 6, "'I Don't Have to Argue My Design—The Visual Speaks for Itself": A Case Study of Mediated Activity in an Introductory Mechanical Engineering Course," Maureen A. Mathison, through analysis of teacher talk and classroom discussions, and student surveys, reports on a qualitative study in which conceptions of writing, initially thought to be congruent, were widely disparate, highlighting the different communication needs of respective disciplines. Sarah A. Bell in chapter 7, "I See What You Mean: Mechanical Engineering Students' Use of Visuals in a Research Paper Assignment," uses the visual as a basis for understanding connections between the visual and the verbal. Her analysis of students' research reports helps her better understand how to integrate the purpose and use of visuals into her teaching.

The last two chapters examine difference as both a barrier and a solution. Chapter 8 highlights the differences between the "hard" and "soft" disciplines as manifested in gender. Awareness of the importance of gender, race, and ethnicity in the core and technical sciences has increased in recent years. Organizations such as the American Association for the Advancement of Science (AAAS) and the National Science Foundation

(NSF) have made a point of emphasizing inclusivity because of the paucity of broader representation across fields. In the chapter "Ideologies of Gender: Culture Clash between the Disciplines," authors April A. Kedrowicz and Julie L. Taylor analyze student feedback in two departments, mechanical and civil and electrical engineering, to determine how student engineers value practices they perceive as "masculine" and devalue the work of the "feminine" writing consultants. In their research, Kedrowicz and Taylor uncover problematic power attitudes about females teaching in male-dominated contexts. In chapter 9 Sundy Watanabe offers a solution to the tensions that arise as interdisciplinary partners grapple with theories, practices, dispositions, and identities. Her chapter, "Intercultural Collaboration: Respect, Relationship, Responsibility, and Reciprocity," offers a novel approach for those thinking about engaging in interdisciplinary work. Employing an Indigenous framework, she examines how difference can be used productively to respect ways of being without the responsibility falling on one group or the other. Finally, Maureen A. Mathison concludes the volume with her chapter "Sojourning, Resistance, and Trust." In this chapter Mathison synthesizes the experiences of the consultants as told through their chapters, and analyzes interviews with participating engineering faculty members with whom the consultants' collaborated to discern areas of success and areas that posed challenges for both groups. In closing she discusses the role of trust—cognitive, affective, and cultural—in establishing more fluid relationships across disciplines.

Programs that integrate writing and speaking into the curriculum have demonstrated measures of success (e.g., Poe, Lerner, and Craig 2010). Data from our own program demonstrate that students' writing improved. To determine if our curricula were supporting student learning, we compared student capstone papers from before we began the collaboration with those of students who had gone through all four years of our interdisciplinary program. On a number of measures students who had completed all the courses in which writing was integrated into engineering statistically wrote better than those who had not (Mathison et al., unpublished). Although our interactions were sometimes rife with tensions, the two groups—writing and engineering—developed a "third culture" that generally placed students at the center of learning.

1
SOJOURNERS AND THIRD CULTURES
Raising Cultural Awareness in Interdisciplinary Programs

Maureen A. Mathison and Mara K. Berkland

Disciplines have long been theorized as cultures with distinct ways of embodying knowledge and approaches to problem-setting and solving (Becher 1981; Becher and Trowler 2001; Snow 1963). Like cultures they embrace patterns of thinking and doing that regulate interactions. In that sense, disciplinary socialization is similar to cultural socialization. "Learning to read and write [and conduct one's self] in academic settings occurs through extended experiences in those settings, by meeting the expectations of those situations, and gaining from the opportunities for participation they offer" (Bazerman et al. 2005, 8). Through disciplinary-specific interactions, students come to learn the "symbols or values [which] represent a constellation of meanings or interpretations that frame a group's understanding of the world and guide social behavior" (Martin and Hecht 1994, 236).

Henry Bauer (1990) articulates the uniqueness of disciplines in terms of their respective practices, including verbal, spatial and mathematical abilities; preferences for theory over experimentation; and even in approaches to pedagogy (216). Elaborating on the cultural variation across disciplines, he explains: "Chemists and historians are not the same sorts of people working at the same sorts of tasks with only the specific objects of work being different, as collectors of coins might differ from collectors of stamps, say; rather, chemists and historians differ as much as do Germans and Frenchmen [*sic*], whose differences of language are part-and-parcel of different intellectual, political, religious, and social habits" (215).

Increasingly faculty are being invited to participate in teaching disciplinary writing in contexts distinct from their own (Dannels and Housley Gaffney 2009; Reave 2004), a situation in which they often find their culture at odds with that of the new discipline; the culture of one discipline does not directly transfer to another. Further, it is difficult to anticipate. As Sue Dinitz et al. (1997) explain, "until faculty actually become involved. . . . they may have difficulty conceptualizing the tensions that may emerge" (41).

Teaching in another discipline creates challenges that have the potential for miscommunication, particularly because of the variation across disciplinary cultures. In the foreword to his book *Cross-Cultural and Intercultural Communication*, William Gudykunst (2003, vii–viii) writes that while most researchers have identified intercultural communication as a field that examines interactions between peoples of different cultures, many are beginning to use more expanded notions of culture, including interactions that Gudykunst labels "intergroup communication"—the communication between "ethnic/racial groups, able-bodied/disabled communication, and intergenerational communication" and so on (vii–viii). In line with Gudykunst, we view interdisciplinary interactions––the "direct, face-to-face encounters between individuals of dissimilar cultural or subcultural backgrounds" (Kim 2001, 141)—as instances of intercultural communication, where disciplines constitute individual cultures.

Applying an intercultural communication and interdisciplinary studies framework lends an invaluable perspective to examine interdisciplinary relationships. The two approaches help illustrate the points we are making and enable us to create an awareness that might assist colleagues address differences as they arise throughout their collaboration. They also advocate for reciprocity of action, ensuring that no one partnering discipline assumes the power and/or responsibility of the interdisciplinary relationship.

In this chapter, we discuss our experiences as interdisciplinary sojourners, which we later use as a context for our unique application of intercultural theory. Next, we demonstrate the connections between disciplines and cultures before establishing the usefulness of an intercultural approach to collaborative work across disciplines. We then identify five intercultural factors that potentially impede interdisciplinary collaborations, and using these factors as a framework, discuss how they potentially impact participants and their goals. Finally, we offer five maxims to raise cultural awareness in order to help avoid and decrease interdisciplinary tension and, ultimately, to help create more successful projects.

THE TURN TOWARD CULTURAL AWARENESS

Historically, the majority of writing instruction for students has occurred as stand-alone courses (e.g., technical writing). In our case, however, the writing faculty were embedded within a mechanical engineering department. Together with engineering faculty, we codesigned a multicourse curriculum that spanned the degree; students performed writing and speaking throughout their design courses, from their first through their last semester (for a description of a model for the engineering curriculum, see Kedrowicz et al. 2005). Members of the WID and engineering team each contributed unique knowledge to the project, considering such things as what types of writing and speaking would enhance the undergraduate curriculum, the point at which those might best facilitate learning, and so forth. Our collaboration is best described as "synthetic interdisciplinarity," where according to Lisa Lattuca (2001), "the contributions or roles of the individual disciplines are still identifiable" (82).

Our approach to teaching and learning was based on situated learning theories, where students learn best in context, that is, in the situation for which the knowledge is relevant and applied (Brown, Seeley, and Duguid 1989; Engeström 1987; Lave and Wenger 1991; Russell 1997). The WID consultants, as they came to be called, were embedded within the culture of mechanical engineering, collaborating with their faculty to design assignments and evaluate student writing; attending classes; providing writing instruction; and working closely with students to provide feedback on and assistance with their writing.

In this new role the consultants became, in Michael Byram's words, "sojourners" (1997, 1). Sojourners are distinct from tourists in that while the tourist temporarily seeks that which has transcended time and change, the sojourner is placed in a situation that indefinitely affects change. While the tourist's life may be "enriched" by the experience, that of the sojourner and the culture in which he or she interacts is forever in transformation because of continuing interaction and influence. Culture becomes fluid rather than static. Byram (1997) explains the role of the sojourner as a catalyst for change: "The experience of the sojourner is one of comparisons, of what is the same or different, but compatible, but also of conflicts and incompatible contrasts . . . the sojourner has the opportunity to learn and be educated, acquiring the capacity to critique and improve their own and others' conditions" (1–2).

The consultants taught in an unfamiliar part of campus in an unfamiliar discipline, with mechanical engineering faculty and students. As a result, they shared more common views about the construction of knowledge and its dissemination than did the consultants. Essentially, the consultants were engaging in an intercultural immersion experience, where one person enters a new culture and learns to adapt. In immersion, intercultural competence—whether rightly or wrongly—generally rests on the ability of the outside member (in our case, consultants) to adapt to his/her new surroundings, initially at a disadvantage as he or she struggles to become communicatively competent, while simultaneously performing in the new culture.

COLLABORATING ACROSS DISCIPLINES

Collaboration among disciplines has increased in the past several decades. A search of the term "collaboration" overwhelms with its millions of hits; it is ubiquitous, a mainstay of almost every discipline as problems become more complex, demanding different types of expertise to solve them. In writing, the concept picked up traction with the publication of Kenneth Bruffee's 1984 article "Collaborative Learning and the 'Conversation of Mankind.'" Drawing from Oakeshott (1962), Bruffee concluded that collaboration was beneficial for students entering the academy because "To think well as individuals, we must learn to think well collectively—that is, we must learn to converse well. The first steps to learning to think better, therefore, are learning to converse better and learning to establish and maintain the sorts of social context, the sorts of community life that foster the sorts of conversation members of the community value" (640). According to Susan McLeod (2001), Writing-across-the-Curriculum was an early adopter of collaborative learning theory. The phrase "Only Connect," taken from E. M. Forster, was an early motto of the writing movement. McLeod explains how those in WAC saw themselves as connecting disciplines, faculty members, and students (viii) to bring about fundamental change in the way writing was positioned in the university. Many tenets of WAC philosophy have been taken up by its sibling, WID.

In his book *Academic Writing Consulting and WAC*, Jeffrey Jablonski (2006) outlines the history and theory of collaboration in WAC/WID. Rather than viewing collaboration as a unified practice, Jablonski carefully maps out three unique models that have guided its practice: the traditional or commonsense collaborative mode, the collaborative philosophy mode, and the consulting model of collaboration. The potential success of the commonsense mode is characterized by the harmonizing aspects of personalities that allow for productivity. Scholarly collaborative writing generally typifies this type of collaboration. Participants in the collaborative philosophy of collaboration tend to engage equally and learn from each other throughout the process, more of a dialogic interplay between partners. Finally, a consulting model (not to be confused with our use of the term) tends to address collaboration through specific channels: workshops, writing center models, writing courses, and expert advice (31–44). According to Jablonski, the first two models tend to take the social interaction of collaboration for granted, while the third, the consulting model, is focused more on the administrative level, without particular attention to the interactions of participants. What all three models have in common, however, are the challenges that collaborative interactions can pose. These range from depth of content knowledge, to ideological differences, to unique pedagogical understandings of effectiveness (e.g., McCarthy and Fishman 1991).

Like WAC/WID, collaboration is the bedrock of contemporary engineering. An abundance of literature exists that elaborates its importance (e.g., Murphy, Davis, and Yurkovich 2009). Engineering students often work in intradisciplinary teams in the classroom to solve bounded engineering problems, but they also collaborate across disciplines to solve complex multidimensional problems (e.g., Terpenny et al. 2006). Professional engineers regularly collaborate with nonengineering experts, for example, in industry, government, and business to provide cleaner air, design safer infrastructures, and develop improved technical devices. And funding agencies today favor interdisciplinary teams because of their synergistic and innovative approaches for solving complex problems (Paletz, Smith-Doerr, and Vardi 2010). Learning to work with others is a timely hallmark of engineering professionalization, one that draws on the ability to be flexible in working with others.

Marie Paretti et al. (2009), however, identify a key difficulty for engineers in collaborating with those in writing. Drawing on Maura Borrego and Lynita Newswander (2008), Paretti et al. believe cognitive flexibility is the most important factor for successful collaboration as it allows for participants to "recognize, value, and adopt multiple epistemological frameworks; that is, they not only recognize other areas of expertise, but they recognize and can adjust to different ways of approaching, studying and disseminating knowledge" (79). Writing, Paretti et al. claim, stands apart from other contexts in which engineers collaborate because unlike other disciplines, writing is often not recognized as a "robust academic discipline" but as a skill lacking intellectual engagement. In other words, the value of writing is not directly viewed as solving complex engineering-related problems. As Bauer (1990) has articulated, practices are regulated by disciplinary members who often hold very highly differentiated perspectives and habits from those outside their own discipline. When they collaborate, differences in epistemologies and ideologies, as well as their concomitant expectations for performance, are highlighted, potentially causing tension.

Much of the discussion on disciplinary tensions has centered on the incompatibility or collision of practices. While first used to explain the conflict between academic and student literacies, Mary Louise Pratt's (1991) notion of "contact zones" has been employed to discern and bridge the differences one encounters across disciplines. These zones are known as the "social spaces where cultures meet, clash, and grapple with each other" (34). Rolf Norgaard (1999), for example, employs Pratt's term to better understand how to rhetorically navigate and negotiate undergraduate disciplinary beliefs about expertise in engineering by "foregrounding the social dimensions" of writing (54). Charlotte Brammer, Nicole Amare, and Kim Sydow Campbell (2008) employ the term to alleviate difference through identifying commonalities, rather than differences, across disciplines. They argue that the "'culture shock' of writing faculty working across the curriculum can be averted by helping writing teachers develop a better understanding of colleagues in other disciplines and their attitudes toward writing" (12).

"Culture shock" remains a central challenge when members of different disciplines collaborate. Anson's (2002) casebook aptly illustrates some of the situations that regularly occur when collaborating in other disciplines, from students (and we believe professors, too) believing in the separation of form and content (Moore Howard 2002), to issues of determining the 'who's, when's and how's' of instruction (Zeleznik et al. 2002). While suggestions of negotiating expertise and finding common ground can attenuate some of the tensions that may occur, few researchers have approached the "clash of cultures" through examining the day-to-day interactions between members of distinct disciplinary cultures as they collaborate.

To date, approaches to improving interdisciplinary collaboration tend to emphasize the relational aspects of interacting and mentoring when engaging in cross-curricular teaching. Deanna Dannels (2005), drawing from Keller's 1983 biography of biologist Barbara McClintock, advocates an approach with a "sincere willingness to 'lean in'" (3), in order to develop a sensibility toward the "organism," which for Dannels would be the disciplinary context and associated practices into which she is immersing herself. J. M. Zeleznik et al. (2002) offer another approach for collaborating. They advocate a method for interdisciplinary mentoring, in which members of the disciplines are "'given the chance to learn from each other as [they] seek a common goal" (231). In this way, dialogue is ongoing, with opportunities to provide cultural instruction regarding assumptions about disciplinary expectations.

Dannels (2005) rightfully articulates the necessity for opening one's self up to another, much like an anthropologist whose goal is to move from an etic to an emic perspective when studying a culture; only in understanding the phenomenon from an insider perspective can one ably discern what and how things mean in particular contexts. "Leaning in" is a starting point for an interdisciplinary collaborative stance but, as we argue, only a starting point, for it addresses WID from the perspective of WID and not the partnering discipline. What is lacking, at least in the discussion, is the reciprocity of commitment and action from the other.

Interdisciplinary mentoring supports the mutual sharing and respect of knowledge, as it emphasizes the more constructive aspects of collaboration. Mentoring tends to focus on the conscious education of interdisciplinary collaborators and entails ongoing negotiation from their respective perspectives. While important, this approach, too, has limitations as it addresses issues as they arise, but does not, in anticipation, proactively address them.

We argue for a more comprehensive approach to interdisciplinary collaboration, one that involves all parties not just leaning out of their disciplinary comfort zones and leaning into the other's, nor focusing primarily on the positive, but coming together to form a "third culture," a "situational subculture wherein temporary behavioral adjustment can be made by the interacting persons as they attempt to reach a mutually agreed upon goal" (Casmir and Asuncion-Lund 1989, 294).[1] A third cultural perspective not only asks participants to understand the relevance of their respective knowledge, practices, and identities as they collaborate, but to also understand that it is not a matter of imposing one on the other; it requires acknowledging and anticipating tensions in advance. By addressing differing expectations, a third culture can be created in which pedagogical needs and concerns take precedence over the cultural moorings of the respective disciplines.

FRAMEWORK FOR ENHANCING INTERDISCIPLINARY INTERACTIONS

As we stated earlier, various current approaches to collaboration in writing can oversimplify the dynamics of interactions, with writing shouldering the responsibility for the collaboration. Jablonski (2006) rightly notes that "the prevailing assumptions and attitudes about collaboration. . . . generally lead to reductive views of collaboration as either a basic people skill or a personal relationship that either works or doesn't work, depending on the social actors' personalities" (31). We challenge these assumptions and attitudes with perspectives from intercultural communication theory, which does not assume innate skills of interaction. Rather intercultural scholars not only acknowledge that there will be tensions, but they also address the potential areas of difference when bringing participants together from unique cultural backgrounds.

Numerous strategies exist outlining ways in which intercultural conflict can be managed or used effectively and how cultures can collaborate to evolve their collaboration and projects to even greater success (for example, see Gudykunst 1998; Gudykunst and Kim 1984; Hampden-Turner and Trompenaars 2000; Ting-Toomey and Oetzel 2001). Ultimately, use of intercultural communication theories and methods can assist interdisciplinary collaborators in being better partners to each other. A first step is to create an awareness of some of the issues that generally arise and to explore their contributing factors. The goal is to shed light on the need and strategies for creating mindful interdisciplinary communication. Tae-Seop Lim (2003) outlines several factors that can interfere with more satisfying interactions: "Lack of knowledge of the other's culture (not on the other's language), ethnocentric attributions, stereotypes, socio-political problems, and unwarranted beliefs of universality are proposed to be some of the major factors causing intercultural miscommunications" (58).

In the following we discuss factors that are critical to consider in enhancing interdisciplinary interactions by employing Lim's (2003) framework, and providing examples from our and others' experiences in interdisciplinary collaborative contexts to illustrate their relevance to writing. Although we outline the factors separately, they are best considered synergistically; that is, they generally underlie the whole of an interaction, with all coming to bear on participants' behavior, with some factors having more impact at times than others on the immediate interaction. We examine each of Lim's factors, beginning with assumptions of universality, as many initially engaging in interdisciplinary collaborations are unaware of its power to undermine common goals.

Assumptions of Universality

Assumptions of universality occur when participants anticipate and expect their ways of being to be similar, if not the same. Many entering interdisciplinary collaborations acknowledge that there are differences between the various participants, but assume an in-group status; that is, they tend to self-identify as academics and scholars who share certain common standards for excellence, particularly in terms of writing. While excellence may be a shared standard, expectations for performance are unique for individual disciplines, as is the manner in which they are valued and achieved; disciplines have unique historical, sociological, and cultural configurations that shape their particular "rules of discourse." Glossing over these differences may initially establish a sense of commonality among collaborators, but ultimately lead to misunderstandings.

For example, in her ethnography of a mechanical engineering classroom, Mathison, chapter 6, shows that though communication was valued by writing researchers and engineering professors alike, it was valued differently. The writing and speaking researchers considered it to be rhetorical and persuasive, that is, an integral part of the design process. They imagined that as students worked on their projects, they would use writing and speaking to communicate to engineering audiences—peer engineers, teacher engineers, and manager engineers. It became apparent after observing the classroom over time that the audience for which communication was important was not the audience they had expected, but that of the machinist. Rather than write an argument for the quality of their design, as would be expected in the home discipline of writing, the engineering students were being taught to rely on the visual to translate a design into a three-dimensional prototype. Said the professor: "Think about how it's [product] gonna be made and give the machinist enough information to show them what you want" (Mathison 2000, 82). Clearly the two disciplines were operating out of different notions of communication, even though the term "communication" was regularly used in classroom discussions by the engineering professor. Until that point, both had presumed that when they talked about helping the students improve their communication skills, they were envisioning and working toward the same goals based on similar expectations. Jessica Leigh Thompson (2009) notes, "a difference in terminology is more than a language-based difference, it illustrates that people from different scholarly backgrounds assign qualitatively different meanings to the same term" (286). They also act differently upon the concepts those terms identify. An example of how, even with the best intentions, assumptions can go awry is illustrated in Sue Dinitz and colleagues' (1997) description of interdisciplinary team teaching, where both instructors addressing the same subject engaged students in accordance with their unique disciplinary interests: "I soon discovered that while faculty did expect their teaching partner to approach the subject matter from a different disciplinary perspective, they did not expect that partner to approach teaching the subject matter differently.

But differences in assumptions about teaching continually emerged in each team: over how to run the classroom, over the relative importance of 'content' and 'process,' over how to approach texts, over teaching the writing process, over how to respond to students, over evaluation" (29–30). Having a common goal does not necessarily translate into a common perspective or practice, nor even a common paradigm from which to proceed.

Lack of Disciplinary Cultural Knowledge of the Other

As reported in Helen Fraser and Andrea Schalley (2009), Michael Wagner described a panacea for interdisciplinary collaborations when he said, "What we need is a series of 'Lonely Planet' guides for the disciplines, in which each provides an overview and background suitable for tourists [and sojourners] from other disciplines" (137). Although such guides do not exist, and could not replace the value of learning disciplinary culture in situ, Wagner's insight draws attention to the lack of shared knowledge regarding disciplinary worldviews and their associated practices. Interdisciplinary participants, says W. H. Newell (1994), "need to develop some feel for the worldview of each discipline, and ultimately they need some awareness of the key assumptions on which those worldviews were predicated" (44).

Tony Becher and Paul Trowler (2001) categorize disciplines primarily based on their epistemological and methodological approaches to the phenomenon they examine. On the one hand, the "Humanities," where writing programs generally reside, is concerned with "particulars, qualities, complication," while, for example, the "technologies" are concerned with "mastery of physical environment." One is considered a "soft-pure" scholarly area, while the other is considered a "hard-applied" scholarly area (36). Engineering, where our sojourners resided, is a discipline that aligns itself more with the sciences in that its conclusions are based on some notion of an objective process. A discipline such as engineering would have a "procedural sense of objectivity," that "focuses on the impersonality of procedure" (MeGill 1994, 10). In other words, there is an accurate technique for doing things, one that is knowledge based, rule governed, and step driven.

In a sense, members who collaborate from unique disciplines are what Susan Star and James Griesemer (1989) termed "marginal people," seeking to straddle multiple worlds. "Marginality," they write, "has been a critical concept for understanding the ways in which the boundaries of social worlds are constructed, and the kinds of navigation and articulation performed by those with multiple memberships" (411). When engaging in interdisciplinary collaboration, group members, like "marginal people," straddle various disciplinary/social worlds at once. Star and Griesemer found in their study of the early years of the Museum of Vertebrate Zoology at the University of California, Berkeley, that its success was due in part to a concept they term "boundary object," an object with which all collaborating members could identity with in some fashion: "Boundary objects are objects which are both plastic enough to adapt to local needs and the constraints of the several parties employing them, yet robust enough to maintain a common identity across sites" (393).

Writing may be considered a type of boundary object in that it is a shared tool, employed across all disciplines; yet because of how it is positioned within individual disciplines, its uses and conventions are often interpreted differently. When collaborating, it takes time to develop a common understanding of the distinct disciplines. Classrooms often become the forum where the differences are exposed as both consultants and the students with whom they engage attempt to discern the other's disciplinary assumptions and norms about writing.

Students are adept at discerning disciplinary difference, in part because of their early exposure to it. It is common for students to ask what an instructor "wants" in fulfilling an assignment, a demonstration of their understanding that they have encountered different disciplinary discourses across the university (McCarthy 1987). Over time students tend to adapt to them, even if they do so unconsciously. In our case, though, students were in their own discipline being instructed by someone who was not; lack of shared knowledge about how the other perceives writing created dissonance between the two. The process seemed more of a nuisance initially for the engineering students, who mirrored the students in Dannels et al.'s study (2003); their research showed that, of the comments in students' reflection logs in a chemical engineering capstone course, 21 percent of the units of analysis indicated that students had a sense that writing tasks were impeding their progress in the course. Said one student, "Let us do more real work," meaning engineering (54). This type of comment was echoed in student evaluations of the mechanical engineering courses in the early years of our collaboration. In the case of WID, writing and speaking may at first be regarded as irrelevant or useless to those in disciplines that tend to rely on other means of communication, such as visuals and numbers as their primary way of knowing, or who may use writing in conjunction with other modes of knowing as Dorothy Winsor's (1994) research on invention demonstrates.

Because academics are deeply rooted in their disciplines, the process of learning another perspective can be unsettling and, at times, threatening. Performing without expertise may be uncomfortable for all involved, as assumptions about knowledge are transgressed, and they are asked to extend their academic reach beyond its current limits.

Ethnocentric Attributions

Ethnocentrism is the negative judgment of another culture based on the standards of one's own culture. Essentially, it is a belief in the superiority of one culture over another (Jandt 2007, 74). Donald Campbell (1969) was one of the first to include ethnocentrism in discussions of interdisciplinarity. However, his concern was directed more toward the ethnocentrism in interdisciplinary infrastructures, how teams would be organized and administered. As interdisciplinarity has become more common, ethnocentrism has been discussed more broadly, to encompass issues of knowledge and value, with some judging their own disciplines as superior: "Teamwork has been compromised by the disdain scientists have for engineers, mathematicians for physicists, pure scientists for applied scientists, physical scientists for social scientists and humanists, and vice versa" (Klein 1990, 127).

Initially it is common for interdisciplinary participants to believe––at least subconsciously—that the understanding and views they bring to a project result in best practices. Each has engaged in long years of training and practice and been recognized for their contributions within a single domain. J. J. Kockelmans (1979) early on likened this specialization to a "deliberate narrowness" (135), which can be advantageously used in collaboration to resolve problems, but which can also be an ethnocentric impediment when the "in-group partisanship in the internal and external relations between academic disciplines, university departments and scientific organizations and institutions" obstructs collaboration (134). Firmly held assumptions about disciplinary worldviews and, therefore, appropriateness of knowledge, skills, and practices over another's create barriers for effective partnership.

By all accounts, engineering and writing may seem at odds as interdisciplinary partners. Their historical roots demonstrate uneasy tensions from very early on. Engineering developed as a practical profession emphasizing the technical over the humanistic (Connors 1982; C. Miller 1979). And the humanistic, with its emphasis on rhetoric, has emphasized the constructive and persuasive value of prose. Patricia Sullivan (2012) colorfully describes this relationship when she says, "indeed, the knot of binaries that ties technical writing and English (often constructed as humanism) has sometimes seemed quite Gordian" (203). More often than not, they seem to operate at cross-purposes. Writing and rhetoric scholars focus on their task—to effectively communicate ideas, beliefs, and propositions—while engineers focus on the task at hand, which is to solve engineering problems. Both rightly live in different conceptual worlds. Louis Bucciarelli (1996) explains how mechanical engineers, for example, focus on the object world and the design process that constitutes it.

The engineering students with whom we worked similarly expected to focus on the design process in order to solve a well-defined problem, and they approached the writing strictly as an explanation of their design. The consultants, on the other hand, believed writing played a more central role throughout the design process. Ethnocentric beliefs on both parts influenced such things as assignment descriptions, expectations for performance, instructor feedback, and evaluation. For example, we perceived writing assignments as an opportunity for students to find their authorial voices within a set of general parameters that pushed their thinking about the design process—to examine, reflect upon, and revise their design, with writing a continual part of the process. The students, however, perceived our general guidelines to open-ended assignments not as tools to assist them but as vague directions that ultimately spoke to the lack of importance of the assignment itself. Ultimately, if there was not a specific, correct answer, the assignments were considered to lack rigor. Thinking the assignments were appropriate, the consultants responded to the student frustration by perceiving them as resistant.

Ethnocentric positions can further entrench already held beliefs when encountering perceived opposition. The danger, as Deborah Andrews' (2003) survey of interdisciplinary writing professors shows, is that ethnocentrism can lead to disparaging views, such as the one below, of those with whom we collaborate: "Students lack adequate motivation. They do not understand that writing up their results is as important as the research itself, that the course will have practical application to their future work. They refuse to put in enough hard work to improve their writing . . . The students have a lifelong dislike for English [*sic*] classes" (447).

Comments such as these can damage relationships. Initially, both parties, even when they understand there are differences may come to the conclusion that the problem lies with the interdisciplinary partner, who myopically is unable to see and act upon things from the "correct" vantage point. Operating solely out of one's own disciplinary lens can diminish its relevance and its potential contribution to the project, while at the same time limit the opportunities for the partnering discipline to contribute. Interdisciplinary partners must learn to trust the other's knowledge, understanding that as both contribute, a different form of knowledge evolves that may resemble that of one discipline more than the other at certain points or that may take on a novel form, new to each.

Stereotypes

Related to the concept of ethnocentrism—which focuses more on the positioning of the self as more accurate, appropriate, and correct in one's assumptions and actions—is the concept of stereotypes. Stereotypes, which are often negative, are "created through socialization, media portrayals, norms, and laws" (Stephan and Stephan 2002, 130). These essentialisms encourage in-group members to make generalizations about others from different ethnic, cultural, or disciplinary identifications (B. Hall 2005). Stereotypes are prevalent, but because they are most often inaccurate they serve more to limit and hinder communication rather than facilitate it.

Unlike intercultural communication, there is little mention of stereotypes in the literature on interdisciplinarity, and yet partners encounter them in their collaborative interactions. Disciplines themselves are often presented stereotypically—through the scholarship that is produced, the objects through which work is accomplished, and even through the dress members wear. In her recent book on academic life, Michele Lamont (2009) addresses stereotypes in interviews with scholars from various representative disciplines. According to one interviewee, an analytic philosopher: "Philosophers are known to be rigorous more than anything else. Right? People in English seem to value a kind of the ability to look at the underside of literary texts and see not so much what they are saying, but what they're not saying. In philosophy that is considered completely useless, whatever. In art history, some want to be on the cutting edge of every last French philosophical movement and be able to bring Lacan, DeLeuze, Baudrillard and Bourdieu into their discussions of the arts" (53–54).

As academics, we tend to hold judgments about others' disciplines. These stereotypes generally reflect a lack of knowledge about the discipline, as well as a lack of contact with a variety of its members in different contexts. A typical stereotype, for example, is that the humanities is more "feminized" and scientific and technical studies is more "masculinized" in its approach to subject matter. Donald McCloskey

(1985), employing Wayne Booth (1974, 17), uses the following binaries to describe both approaches: "value/fact, opinion/truth, subjective/objective, normative/positive, intuitive/rigorous, vague/precise, words/things, feeling/cognition, soft/hard, yin/yang" (McCloskey 1985, 42). Rigid in their determination, these stereotypes leave no option for other ways of approaching subject matter. Without exposure and experience to disciplinary cultures, inaccurate perspectives can develop that undergird expectations and guide interactions.

One prominent example of how stereotypical expectations influence communication came during the first year of our collaboration. One of the researchers received a memo from a student that began "Since the dawn of time." When the researcher returned the memo and asked the student to revise it, explaining that the document was inappropriate for engineering, the student responded, "Well, you're in the humanities and you like that fluffy stuff." The response communicated a previously held expectation about what a humanities professor would require from the writing task, while at the same time communicated that it was not valued within the engineering culture ("fluffy"). It may also have indicated the student did not have confidence that a member of the humanities could understand technical information, an issue that emerged over time.

Dona Reese and Mary-Ann Sontag (2001) further clarify the complexity of stereotypes in explaining their experience with interdisciplinary collaborative work. Examining patient experiences with chaplains, social workers, and nurses, they explain that there is sometimes an unwillingness to work with certain team members because of misperceptions about the limited skills or benefits the discipline can provide. We found in our experience that in the early years, our engineering collaborators often saw our skills as being limited to "English only"—grammar, mechanics, syntax. They could not fathom that we could bring more than sentence-level editing skills to correct student writing, a blow to our sense of mission and selves. As our interactions with them increased, they began to progressively understand that our expertise helped us situate, explain, and teach writing in an engineering context in ways they had not considered. For example, they learned about engineering genres and their standardized terms, implementing them across courses and levels. What once had been erroneously labeled a "memorandum" in a laboratory course was now accurately labeled a lab report. The discussion of poor writing as grammar skills became more complex as we spoke of organization, technical style, and use of jargon. Eventually, they came to understand that we provided them the rhetorical knowledge that would make their documents more professional and compelling—significantly more than the wordsmithing skills that they had originally thought we could contribute.

Sociopolitical Issues

At its core, interdisciplinary collaboration is sociopolitical. Peter Weingart and Nico Stehr (2000) write that "disciplines are not only intellectual but also social structures, organizations made up of human beings with vested interests based on investments, acquired reputations, and established social networks that shape and bias their views on the relative importance of knowledge" (xi). Realistically some disciplines, and individuals within those disciplines, are perceived as possessing more value than others. A hierarchy of value exists in higher education.

As a result, power differentials exist at the level of authority and expertise. Laura Ellen Hirschfield (2011) explains a useful distinction between the two when she writes: "Loosely defined, authority is the power to make and enforce decisions, while expertise is a series of mastered skills with increasing levels of difficulty in specific areas of functioning (Turner 2001; Ericsson 2006). Despite the presumption that people with authority are the experts, as well as the opposing norm, that higher levels of expertise accompany authority, this relationship does not always hold constant (Clancey 2006)" (3–4).

To maximize interdisciplinary collaboration, experts from participating areas share their mastery of knowledge with equal authority. Yet, as Hirschfield points out, this is not always the case. In collaborating with engineering, two factors may unconsciously influence the exercise of authority and expertise. Because science, technology, engineering and math (STEM) disciplines are more heavily funded in today's climate, they tend to be perceived as having more value to society. Because they are male dominated and with little diversity, they are more insulated in their taken-for-granted social interactions. The privilege and power STEM disciplines hold often create a disjuncture in authority and expertise between them and writing.

As a discipline, writing tends to be dominated by females (at present, mostly at the assistant and associate levels). As a practice, writing is deemed necessary but not taught to the extent it is claimed to be valued (within and across the curriculum). Combined, the two can place writing consultants in a lower sociopolitical status with less leveraging power. From a general perspective, how is time allocated in class for a valued practice that is considered necessary, but ancillary, to the prime objective of a course? Second, how might material be delivered by female consultants, who generally receive different treatment in the classroom than their male counterparts (Carli 2001; Chesler and Young 2007)?

At a very basic level, questions related to time and the distribution of content and its delivery can pose critical challenges for WID. Who should teach what, when, and for what period of time? In her survey of cooperative writing programs, Andrews (2003) found that the third-most-common struggle of WID professors was that the cooperative discipline did not allot enough time to the study of writing. While outwardly a question of pedagogy, time-related issues, coupled with content, can create tensions about the value associated with the different types of expertise.

Related to the larger issue of curriculum design was the day-to-day concern over specific class sessions. Because writing lectures and assignments were integrated into the engineering classes, there were ongoing discussions about how much time a particular lecture might "take" or "need." Tensions arose when one collaborator's lecture ran too long and cut time from the other collaborator's lecture. Such unplanned uses of time, which are common in the classroom because of student questions or discussion threads, were often misconstrued initially to be symbolic of the lack of value that one discipline placed on the other. Access to students and control over the flow of knowledge were together considered a valuable commodity by all partners, who felt time directed toward one area of disciplinary development meant less time directed toward their own.

Equally important, the writing consultants were usually female, which awkwardly positioned them as they taught in a male-dominated discipline. To begin, they were considered disciplinary out-group members, and, yet, as instructors they were in positions of authority in the classroom, teaching students about the rhetoric of engineering and its applications. This double bind was confusing for both students and consultants and complicated by gender issues.

Historically, females have been excluded in engineering. As late as 1955 the dean of engineering at Penn State University, Eric Walker, wrote that most women did not have the "basic capabilities" to successfully work in the technical sciences (Bix 2010). Such stereotypes persist today. According to Catherine Hill, Christianne Corbett, and Andresse St. Rose (2010), females are still considered to be less qualified to work in math and the sciences. Their American Association of University Women Report (AAUW) explains that "although largely unspoken, negative stereotypes about women and girls in STEM are very much alive" (38). Many females have responded to this stereotype, staying away from these fields. In 1966, 0.4 of engineers were female; today, that number has increased to 19.5 percent (9). While there is this higher proportion, women remain a minority. Writing consultants in engineering were part of that minority, with all the stereotypes that have pervaded engineering culture. How could they possibly teach writing in an engineering context when they did not have the ability to "get" the science?

It is critical to be aware of power differentials. Issues of allocation of time—and certainly issues of race, ethnicity, class, gender, rank, age, and resources—all impact interdisciplinary interactions. In our experience, the WID consultants initially tended to live on the terms of the engineering partner, not necessarily intentionally but by default. Although they had expertise in writing, they were not granted full authority. They were from a lesser discipline and were of lesser ability in STEM, both of which compromised their position. As a result, interactions surrounding issues of power occurred on a daily basis with both engineering faculty and WID consultants, as they learned to compromise and work through the tensions in order to better attain the goals of the collaboration.

Interdisciplinary collaboration is negotiated; at the heart of interdisciplinary interaction is the ability to minimize the cost to pedagogical interests and identity.

MAXIMS TOWARD CREATING A THIRD CULTURE

Interdisciplinary collaboration entails that participants recognize that even when sharing a common goal, it is approached from highly unique and, most likely, highly different perspectives. Rainer Bromme (2000) writes, "As everyday perception of facts and events depends on the categories we bring to a certain situation, the question of how mutual comprehension in the case of different perspectives is possible arises" (118). How then is it that interdisciplinary partners can contribute differing knowledges, beliefs, and practices, yet achieve a coherent and shared view of their goals? Toward that end, we submit five maxims that offer a sensibility toward interactions and that can also help to improve communication—necessary for creating a third culture dynamic:

> 1. *Having a common goal does not necessarily translate into a common perspective or practice, nor even a common paradigm from which to proceed.*

Intercultural scholars advise collaborative partners to accept that there will be differences and that these different perspectives can quickly lead to tensions and conflict (Ting-Toomey and Oetzel 2002). Anticipating that there will be conflicts is important, as it allows participants to proactively engage in dialogue about expectations and assumptions in order to develop conflict strategies of which the entire team is aware and to which it can commit.

> 2. *Performing without expertise may be uncomfortable for all involved, as assumptions about knowledge are transgressed, and scholars are asked to extend their academic reach beyond its current limits.*

Interdisciplinary perspectives may initially seem alien and at times even inferior in comparison to one's own. In some cases, new perspectives may even threaten one's own disciplinary viewpoints. In either scenario, interdisciplinary partners can experience discomfort or mistrust. When conflicts arise, the most common responses—competing, avoiding discussion, or accommodating a dominant perspective—might seem the easier path, as they do not require an examination of one's individual disciplinary assumptions. However, these strategies are relatively ineffective if the goal is to create a project that embraces the strengths of all disciplines involved (Coleman and Raider 2006; Gudykunst 1998).

Comparatively, compromising and collaboration strategies require significantly more time and intellectual and emotional effort than ignoring the differences. Learning a new perspective or adapting one's disciplinary practices while working in a collaborative environment is taxing. Such intense exploration may also cause participants to question one's own discipline or propose scholarly transformations that may initially put one on the margins of one's own discipline. The process of examining one's previously held assumptions, though uncomfortable, is rewarding as it often results in better projects and stronger interdisciplinary relationships (Coleman and Raider 2006; Hampden-Turner and Trompenaars 2000).

> 3. *Operating solely out of one's disciplinary lens can diminish the relevance and its potential contribution to the project, while at the same time limit the opportunities for the partnering discipline to contribute. Interdisciplinary partners must learn to trust the other's knowledge, understanding that as both contribute, a different form of knowledge evolves that may resemble that of one discipline more than the other at certain points, or that may take on a novel form, new to each.*

Compromise and collaboration require that we embrace other perspectives and step out of what we "know." Drawing primarily from the collaborative negotiation model (Raider, Coleman, and Gerson 2006), a collaborative approach is one in which both groups attempt to incorporate their goals and find common methods through which to meet these ends (Coleman and Raider 2006). This approach allows both groups to be invested in the goals they work toward attaining. This also allows the groups to establish the normative processes and guidelines that will be employed toward the completion of goals.

Within the process of collaboration and compromise, there are simple steps that can be employed to make interdisciplinary interactions more fruitful. First, it is imperative to negotiate and record the goals the participants have for a program, project, assignment, or presentation (Coleman and Raider 2006; B. Hall 2005). Stating the goals forces a group to prioritize and focus their activities in a way that puts assumptions up for negotiation and potential incorporation into the project. Second, because conflicts are inevitable in such interactions, the maintenance of a positive group climate is paramount to retaining participants' commitment. If they are not committed, they are more likely to engage in less productive conflict strategies that negatively impact the group environment.

> 4. *Stereotypes are prevalent and limiting, difficult to discuss because of their sensitive nature and ability to insult.*

While the collaborating partner is in a discipline unique from one's own he/she should be respected as someone with whom to connect, and not alienate. Such stances help to create a consciousness toward mutual communication.

While many theoretical approaches attempt to examine issues related to intercultural communication, none solely can respond to all. One approach to eradicating stereotyping is that suggested by Judith Martin and Thomas Nakayama (1999), who remind us that "people are both group members and individuals and intercultural communication is characterized by both" (15). That is, rather than address someone as a predefined member of a discipline, recognize in the moment that the participant is also an individual. Essentially, it means "letting go of the more rigid kinds of knowledge that we have about others and entering into more uncertain ways of knowing about others" (18). In letting go, one is better able to respond to engage interdisciplinary partners not as members of a group, but as individuals.

> 5. *Interdisciplinary collaboration is negotiated; at the heart of interdisciplinary interaction is the ability to minimize the cost to pedagogical interests and identity.*

Successful interdisciplinary projects require that all participants have equal status and decision-making power (Coleman and Raider 2006; B. Hall 2005); if not, an authentic interdisciplinary project transforms into an adaptive one, where one discipline dominates and the other follows. Disciplines may be perceived as having more authority because of their histories or current social standing, or because of expectations related to gender (or race or ethnicity, or sexual identity). To avoid inequities, interdisciplinary members should vary the contexts for meetings when appropriate. This approach helps empower representatives from the different disciplines, who may unconsciously be involved in "my turf / their turf" dominance, and it gives every participant the opportunity to learn to feel comfortable and immersed in each other's presence and space. It also allows participants to experience the hard/soft science dichotomy, which highlights issues of gender. Further, consistent efforts to communicate and metacommunicate can make the process smoother. Interdisciplinary partners should feel comfortable and encouraged to clarify misunderstandings and communication styles in order to facilitate awareness through attentiveness to each other's needs. "Mindfulness can be practiced and reflected through a deep state of listening without judgment. It also means exercising creativity and flexibility to seek out a "third culture" approach to bridge the cultural differences" (Ting-Toomey 2007, 259). The commitment to open communication helps participants recognize and clarify assumptions so that goals can be better understood and negotiated by the entire group.

Each of these maxims derives from the sojourner experience, learning in situ how to engage another disciplinary culture, whose practices represent difference. In our case, we, as WID consultants, and the engineering and science faculties with whom we worked, also found that collaborative, cross-cultural approaches to curriculum development were much more effective than having us try to adapt or incorporate our expectations into their previously set assignments and lectures or theirs into ours. In reflection, we noted that the approach was successful because the employ of a cross-cultural perspective did not privilege one discipline's knowledge or perspective over another's, even though we had been invited to work with members of another discipline within their programs. Through negotiation and establishment of joint goals, we were able to maintain a climate of equal status and value for each team member's contributions. We created assignments together, worked on lecture goals and connections as a group, and, ultimately, though the process was time consuming, the project was generally successful.

The implementation of intercultural conflict prevention strategies greatly helped to identify and negotiate assumptions and incorporate the strengths of our different perspectives and knowledge to the benefit of students. As Ting-Toomey (2003) explains, adapting one's goals and behavior to the situation "signals our mindful awareness of the other person's perspectives, interests, and/or goals, as well as our willingness to modify our own interests or goals to adapt to the conflict situation" (334). It does not, however, mean placing one's own knowledge and practices second, but rather placing them alongside those of the participating discipline. This process opens the door to the creation of new goals and behaviors that combine the best contributions of the participants, essentially, a third culture.

CONCLUSION

In collaborative efforts a multiplicity of cultures with different knowledges, perspectives, and approaches to problem setting and solving can, in combination, lead to novel insights and solutions. To provide a more inclusive view of knowledge, the base-cultural perspective of all participants must be taken into consideration. Writes the New London Group (1996), "The human mind is embodied, situated, and social. That is, human knowledge is initially developed as part and parcel of collaborative interactions with others of diverse skills, backgrounds, and perspectives joined together in a particular epistemic community, that is, a community of learners engaged in common practices centered on a specific (historically and socially constituted) domain of knowledge" (82). Thus, when we work in or with other disciplines, especially in the cases of writing-in-the-disciplines, we move from our own "historically and socially constituted domain of knowledge" to another, where that domain itself encounters a newness of culture—ours.

In the case of WID reciprocity is a crucial, yet challenging task. Despite the fact that this case study situation is relatively unique, the ideas of interdisciplinary travel and collaboration are not. Writing-in-the-discipline educators work on assignments for diverse courses, participate in workshops in a diversity of disciplines, and team-teach in a variety of settings, all circumstances that put them in the position to encounter these issues. In each of these circumstances, the educator takes on the role of the sojourner, residing elsewhere, if only briefly. This immersion process has the tendency to construct difference as a lack of knowing. Because the sojourner is surrounded by people within a different system whose dominant assumptions and communication patterns tend to be more similar and cohesive, he or she is often regarded as the person who needs to change and learn. Many times the sojourner falls in line with this way of thinking. However, for one person or small group of sojourners to merely change and adapt undermines the purpose of interdisciplinary interaction. The goal is to share knowledge to create a new way of thinking that is better than either group could have conceived of alone.

Dedication to an awareness and respect for difference bring about the collaboration and power equity that are key to the sharing of ideas. Within interdisciplinary collaboration, new rules, ideas, and communication patterns potentially emerge. This synergy can only happen, however, if collaborative participants work to recognize they are different, understand what those differences are, and set aside the biases or power struggles that can appear as a result of those differences. Once these changes occur, the group can come together and create a third culture dynamic unique to their situation, knowledge, and needs.

As Suman Lee (2006) explains, "In the conjoining of their separate cultures, a third culture, more inclusive than the original ones, is created, which both of them now share. Third culture is not merely the result of the fusion of the two or more separate entities, but also the product of the harmonization of composite parts into a coherent whole" (294). The creation of the third culture allows two ways of thinking that may have seemed irreconcilable to conjoin in such a way that relevant strengths of each become compatible and, more important, pedagogically integrated to better address student needs.

NOTE

1. Originally, the term "third culture" was coined to refer to expatriate children whose families lived outside of the United States for lengthy periods (Useem and Downie 1976). Ruth Hill Useem (1999) later argued that these people integrate elements of their home culture with and their various cultures of residence into a third, different, and distinct culture and may no longer feel at home in any one specific culture.

2
PROFESSORS DESIGNING ASSIGNMENTS AS RELATIONAL ACTIVITY
A Baseline for Connecting Thinking, Learning, and Writing

Maureen A. Mathison and Linn K. Bekins

Writing, recognized as part of an activity system, is one of many tools used to aid thinking and motivate people to action, and it is a means for communicating messages that others will use for some of those same purposes (Bazerman 1994; Russell 1997; Prior 1998; Spinuzzi 2003). This current view has shifted the emphasis away from scribal writing, seen solely as a set of "correct" features to a more fluid and rhetorical view, a nexus where the cultural, social, and cognitive coalesce as writers identify and negotiate the reasons for which and the audiences for whom they ultimately will produce texts (Downs, chapter 3). In higher education, writing plays an important role for learning about and performing in a discipline. Through exposure to the tenets and conventions of a discipline, students learn to perform knowledge and apply it to particular situations (Ackerman 1991; Berkenkotter and Huckin 1995; Walvoord and McCarthy 1990; Winsor 2013). They are learning the "Discourses" of their majors (Gee 1989); they learn how to communicate and comport oneself as a genuine member.

In this chapter, we examine how three professors in the College of Engineering purposely assigned writing in their courses and ask:

1. If writing is a tool within a web of activity for learning, then what goals do professors have for student learning?
2. How and where do writing assignments fit into the course?

Many studies have examined how students respond to assignments, but as Anne Beaufort explains, "the design of writing assignments is under-researched and under-discussed" (2012, 477). Our focus is on the activity of the classroom as it is embodied in the goals of the syllabus and enacted through assignments. Syllabi and assignments provide students

DOI: 10.7330/9781607328032.c002

a type of social and intellectual guide to disciplinary knowledge. Syllabi serve as signposts for learning, and assignments serve to reinforce and extend it. "Instructors' writing assignments," explains Dan Melzer, "say a great deal about their goals and values, as well as the goals and values of their discipline" (2009, 240). While many instructors claim they support student writing in their discipline (and they do), just as many do not understand the multiplicity of writing, even in the same discipline. They often overlook that while their responsibility is to support learning of disciplinary material, writing is integral to that material in ways that extend beyond the page. Writing assignments in fact mediate disciplinary knowledge and disciplinary practice.

Central to this chapter is the belief that disciplinary knowledge is mediated; that is, in the classroom students are introduced to the relevant symbolic forms and tools of a community that allow for the construction of knowledge and for the imbuing and sharing of its meaning. These forms and tools are multiple and overlapping; they may take the form of the verbal, visual, or mathematical. They may be print or digital. In particular, we examine writing as a tool that does not single-handedly move students toward appropriating disciplinary identity. We argue that writing, like other forms of social life, is relational (Russell 2002b). This is to say that writing is a cultural tool that may indirectly facilitate discursive ways of "behaving, interacting, valuing, thinking, believing, speaking, and often reading and writing" (Gee 1996, vii) as it is placed within a web of instruction.

In the introduction Mathison discusses the role of genre as a resource for the activities that undergird the practices of institutions and organizations. As Carolyn Miller (1984) has established, genres are "typified social actions"; that is, they include particular types of information in a normalized and anticipated order, are directed to specific audiences, and utilize distinctive tones for particular uses, occasions, and purposes. Genres develop and transform over time in response to various exigencies, whether writing for personal, social, or professional reasons (Bawarshi 2003; Devitt 2004). More recently, David Russell (2010) has integrated conceptions of genre and activity theory, articulating the important relationship between the two, especially as students learn to write in specific disciplines. What is most compelling about his argument is the idea that contexts are connected; writing in one context supports writing in other contexts and for potentially different purposes.

If we consider this proposition wholeheartedly, then students writing in their majors—in this case, engineering—will be given opportunities to anticipate multiple contexts: people writing to communicate messages foresee the occasions for which that writing will be useful, whether it be to consider, analyze, or report information and alternatives, or to advance a particular claim. Yet, writing in the classroom often times ignores a rhetorical approach and continues to be more evaluative of content knowledge and grammar rather than an instructional tool for wider purposes.

One way to better understand writing within a web of instruction is to examine how writing is articulated by professors as they design courses and instruct students. To gain insight into writing in an engineering context, we examined writing as it was positioned among other activities in the classroom, focusing on how they worked relationally to potentially facilitate disciplinary ways of knowing. Through analysis of interviews with engineering professors about the design of their courses (as interpreted through their syllabi), we provide three examples of classrooms and their mediated activities. The research was conducted shortly before the onset of our collaboration with engineering. Findings show the multiple ways writing had been integrated into courses, serving as an introduction to the various writing cultures of engineering classrooms.

In the following sections, we outline our methodology for this study, address our research questions, and discuss how writing assignments serve as social and intellectual guides for disciplinary communication. We then present findings from interviews with three engineering professors, each illustrating unique ways writing activities were mediated through their understanding of and disposition toward classroom material. Finally, we discuss implications of these results for interdisciplinary researchers and practitioners with an interest in integrating writing assignments into the teaching of engineering.

METHODOLOGICAL APPROACH

The three professors in this study were recruited from a current list of professors who taught undergraduate courses in the Writing Emphasis / Writing Intensive Program at our institution, a large, R1 university. These courses follow certain criteria, among them that professors must offer multiple opportunities for students to write, receive feedback, and revise in various genres, and to provide instruction in the classroom in order to help students understand the purposes for writing and how to best appropriate them.

There were approximately fifty professors in the sciences, social sciences, and humanities to whom letters were sent. Of the forty who responded, twenty-four professors volunteered. Three of the twenty-four professors were in the College of Engineering and participated in this study. We selected these three for this particular analysis to better understand how professors in engineering incorporate writing into their classrooms; others in the corpus of twenty-four respondents represented other disciplinary cultures and were not relevant for this specific analysis. Thus, our sample, though small, was purposeful.

Similar to Chris Anson and Deanna Dannels' (2009) methodology of formative assessment in departments, we interviewed professors and collected their course materials. Combined, the interviews and materials presented us a snapshot of the positioning of writing in their individual courses. In order to construct a description of the writing completed in the courses, the interviews we conducted were intended to be flexible, with planned questions as well as emergent ones (Creswell 2014). Interviews lasted approximately sixty minutes; interviewers asked professors to explain their course and the role of writing in achieving their pedagogical goals. In addition, professors discussed their views on writing in their respective specialization areas. Interviews were recorded and transcribed. Independently, we inductively analyzed the transcripts for thematic categories (Miles and Huberman 1994), and as we read we identified portions of the transcripts that recurrently mentioned specific concepts about teaching. We then discussed our independent findings and collaboratively arrived at three categories: (1) learning engineering content, (2) learning to write, and (3) writing to learn. The first category focuses on discussions of teaching about disciplinary content; the second focuses on what it means to write as an engineer; and the third, "writing to learn," focuses on how writing can facilitate the learning of disciplinary content. To further avoid predetermined outcomes, we triangulated the results of our analyses of the syllabi, interviews, and course assignments.

Discourse segments in which professors emphasized learning disciplinary content were coded as "engineering content." The following is an example of a professor explaining his approach to testing for content: "So, what I do is give four one-hour exams, and they'll study a lot for those exams than they would ever do [for] homework in my experience. What I do with the exams is first of all tell them what will be on the exam. For example, I'll say you will have four problems on the exam. One problem will consist of circuit elements . . . and you will have to calculate any voltage within that circuit." This helped us identify how engineering content was positioned in the classroom.

Discourse segments in which professors emphasized learning to write, were coded as "learning to write." Below is an example of a professor explaining his approach to teaching disciplinary writing in class: "I require them to have some sort of an introduction. So, I just tell them to write an introduction that sets the stage for what's in the report. It's a statement and a subject and the purpose and the scope and a few things like that, but I can't hope to really teach them how to write an introduction, but they need to know that a report has an introduction." This helped us understand how individual professors in engineering approached the teaching of disciplinary writing in the course.

A third category, "writing to learn," emerged. In this category, professors made mention of engineering and writing in the same discursive segment: "So [students] have to understand principles, principles of chemistry, principles of physics, and so I think they need to understand these things. . . . and I think by writing these things down, by going through and writing out definitions or explaining things, that they actually learn those principles better." This helped us understand how these professors used writing for engaging with engineering content. Additionally, we culled from the interviews and course materials the types of writing instruction typically students received in each classroom (see table 2.1).

Interview transcripts were inductively analyzed in conjunction with course materials through a process of constant-comparison (Glaser and Straus 1967). This process gave us insight into how each professor employed writing in the classroom as a tool for thinking, learning, and communicating about the discipline.

THREE CLASSROOMS, THREE APPROACHES TO WRITING AS AN ACTIVITY

The Chemical Engineering Classroom

The senior chemical engineering course presented writing as an unproblematic transmission of information. In the words of the instructor, the purpose of writing was to "keep a record of how you solve the problem . . . to communicate it to your manager so that they can decide what to do with it [lab data]." The course was a "sort of catch-all of things that [students] encounter in the real world . . . and writing is a part of that." As such, students were asked to run experiments; write up their objectives, methods, and results; and then describe the theory behind their work. The course enrollment was relatively small (thirty-five students), and while most of the experiments were collaborative, all students wrote up their work individually.

Table 2.1. Instructional Resources for Student Writing

Course	Average Number of Students	Writing Instruction Resources
Chemical engineering	35	One class period devoted to teaching report genre structure; one class period focused on grammar and stylistics for engineers; student given models of report writing and problem statements; guidebook provided on technical elements, graphics, and grammar;. feedback provided by expert in English.
Electrical engineering	65	No direct instruction on writing; students shown engineering journal articles; students given models of report writing; handout provided with report requirements; feedback provided by engineering teaching assistants.
Materials science and engineering	7	One class period devoted to writing expectations; minilesson given on writing before each lab; draft provided to professor for feedback; instructor conferences with students on drafts;. feedback provided by professor.

According to the professor, the course was intended to professionalize students by offering "real-world" experiments that do not always work out as anticipated, presenting ethical cases for students to consider, teaching computer programs for the specific applications of the discipline, and requiring students to report on their experiments both orally and in writing, just as one might in the workplace. Writing assignments were simply conceived as "a permanent record of what went on."

While writing practices are important means by which to facilitate students' learning of the types of thinking and rhetorical conventions that are associated with their chosen disciplines, this professor saw writing as formulaic, defined primarily by correct grammar usage, and as a static communication vehicle with little emphasis on rhetorical elements, such as audience, purpose, and context. While this professor required students to articulate disciplinary information through their written work, he did not make explicit what it means to write like an engineer. The disciplinary knowledge was treated separately from the communications. "It's largely a matter of communicating what you've done," explained the instructor. As such the written reports received two grades: one from the engineering professor on the technical quality and content and the other from an English specialist on the quality of the writing (e.g., two grammar errors resulted in a rewrite of the whole report). While both content and form were valued, they were treated separately.

Several handouts emphasizing a clear and concise writing style and the report format were distributed to students to help guide them. "Simple, direct English is the best salesman for your ideas," stated one handout. But what does this mean in the context of writing up lab results? According to one handout, "'clear and concise writing can be attained by achieving a low 'Fog Index:'" "Fog Index = 0.4 (No. of words per sentence + No. of polysyllabic words per sentence). This is roughly equivalent to the number of years' schooling a reader must have had to be able to comprehend the written matter on first reading . . . If you write anything with a F.I. above 12–14, you severely limit your audience. If you really have something to say, you can say it with a Fog Index of 10 or less." Clearly, this view of writing is static and objective; that is to say, it separates knowledge and its articulation from social practice. Although writing weighed considerably in the talk about training students for the workplace, its instantiation in the classroom focused on precision and correctness for the delivery of engineering content.

The Materials Science and Engineering Classroom

The materials science and engineering course was a required laboratory course offered to second-year students who had completed two prerequisite lecture courses. This course introduced them to core principles and concepts in the field, which draws heavily from chemistry. According to the professor, students were expected to become familiar with laboratory equipment, to apply theory to practice, and to learn to write formal reports. The laboratory supplied an integral context for the course, where theory and practice converged, with multiple writing assignments incorporated into it to leverage learning. The professor elaborated: "if they have to explain some of this theory, some of the principles behind the theories, and they do it in a laboratory then they learn a little better."

According to this professor, the descriptive nature of materials science and engineering (as opposed to other analytical and largely math-driven engineering fields, such as mechanical or civil engineering) also made communication skills particularly important for students. He stated that students must understand and be able to explain particular conceptual and theoretical principles in order to do well not only in the course but also in the discipline. Writing, he believed, plays a role in the learning of content: "by writing these things down, by going through and writing out definitions or explaining things, they actually learn those principles better."

In a ten-week period, the students completed seven experiments. According to the professor, he had designed writing assignments that are realistic for the workplace. "This is something an engineer would do," he explained. For every experiment, a report, either in memo or report form, was individually assigned. The purpose of each writing assignment was "to tell people what [you] do or what [you] want to do." Initially class time was spent discussing the purposes and types of reports, what information goes into each section of a report, how to express mathematical principles in writing (e.g., write 0.1 instead of .1), how to use advanced features of a word processor, and so on. By beginning each class with a discussion of genre and disciplinary conventions, the professor strategically positioned writing as integral to engineering practice, and not as an add-on. Students spent the majority of class time running the experiments and then made appointments with the professor to individually review drafts and discuss graded work. Through explicit guided instruction and feedback, the professor offered students knowledge about entry into the discipline. He reported typically spending up to forty-five minutes with each student for each writing assignment, supporting students' growth as engineers. The small class size (only four to eight students enrolled per term), coupled with abundant individual attention to students' writing, helped students improve their communication while learning content.

Yet despite the significant emphasis on and support given to writing, the professor noted that students complained they had to write like him to get a good grade. Perhaps this is realistic, given the professor's remark "I try to explain to [students] why I would reorganize things this way, or why this doesn't make sense, or why it would sound better this other way." Given the genre focus in the writing activities and the extensive feedback, we surmise that while students might have grumbled, the professor's input was meant to help them learn to rhetorically understand what it means to write as an engineer for an audience of materials science professionals.

The Electrical Engineering Classroom

The first-year electrical engineering course we studied was offered as the second of three consecutive courses that served as "fundamental" introductions to electric circuit theory and practice. The course enrollment fluctuated between 60 and 100 students and was required for electrical and computer engineering majors. This particular course was neatly organized into four units, with each unit consisting of a set of learning objectives, a study guide, classroom discussions, and a one-hour exam. The "Course Procedure" as the professor titled the syllabus, began with a lengthy description of the professor's belief that through practice, learning occurs, and in order to learn, one must get expert feedback on one's work. Each unit was designed to help students practice "the skills that they need to learn to be able to solve problems," which included performing mathematical operations, applying knowledge of theorems, and so on. Classroom discussions were generally centered on problem solving and supported through laboratory assignments, where students solved authentic problems and recorded their data from the procedure. Students were given access to previous exams to think through problems and also to teaching assistants, who helped them work through exams when they were too challenging.

In brief, this professor's pedagogy created a "miniature professional experience" by (a) presenting students with actual workplace problems and then asking them to solve and communicate their solutions; (b) "coaching" students to practice solving the problems presented in class; and (c) providing students examples of "typical articles in engineering journals to follow" as well as feedback on their writing. Each activity introduced students to the sociocultural activities inherent to electrical engineering and was designed to enculturate them into the discipline.

The writing assignment sequence began in the lab, where students were required to keep a lab notebook that detailed their process and findings. The basic guideline of a good notebook, according to the professor, was "whether or not they have recorded sufficiently what they did . . . to go down and reproduce your work." Drawing from their notebooks, students produced a formal report, which was based on the data and diagrams, but was expanded and revised to include the conventions of a report. The professor explained to them that this included an introduction and conclusion written in concisely composed "standard English."

The formal report assignment was designed to model the processes of "real" engineering. That is, the professor presented the students with an actual workplace problem that needed to be solved; he then asked students to choose a mathematical representation to solve the problem and to design a device based on the mathematical model. He then required them to construct and test the device; and finally, he asked for a formal written report describing the work to their engineering peers. In this manner, students "learn about the real application of the principles to design the devices" in that they may have "to explain why the calculated and the measured results are not the same."

Expectations for the written formal report this professor explained were threefold: (1) to communicate data from the lab notebook and effectively communicate it to their peers, (2) to learn about their technical work through the process of writing it down, and (3) to reflect on a design and account for results. In this assignment, writing was primarily viewed as a transactional tool to communicate data in a format acceptable to the engineering community. Journal articles, as well as the course textbook, served as disciplinary models, but beyond this, little specific instruction was given. Using these materials, students learned by doing, through trial and error, without the benefit of direct rhetorical instruction. They were engaging in what Winsor (1996) refers to as "appropriate rhetorical behavior even before they [understand] it" (42). Although this laissez-faire model can hinder students' progress in producing texts acceptable to an engineering community because of its lack of transparency, students can potentially learn if they persist.

Teaching students to write necessarily involves both a cognitive and a social dimension. This professor viewed teaching as a reflective process that may improve students' ability to learn concepts by explaining their activities in writing. This can be seen in the assignments that asked students to develop a rationale for representing problems, propose solutions based on mathematical principles, and communicate criteria to evaluate potential solutions. The lab notebook functioned as both a data source and a reflection support tool from which they could construct their text with highly specified generic sections (introduction, methods, results, discussion), which are defined by (and used in) engineering communications. In this sense, students were being taught to make arguments stemming from asynchronous classroom activities, some of which were more obviously connected than others and with some receiving more pedagogical support than others.

In each of these classrooms, we have seen how writing is viewed as a tool to be employed to engage students in thinking about issues and problems in their field as well as used in relation to other tools to support specific disciplinary purposes. However, for the professor in materials science and electrical engineering, writing was to varying degrees an embodied activity, inherently relational: "it takes place either in a collective—i.e., jointly with other people—or in a situation which the subject deals directly with the surrounding world of objects—e.g., at the potter's wheel or the writer's desk . . . the human individual's activity is a system in the system of social relations. It does not exist without these relations" (Leont'ev 1981, 46–47). Through such a theoretical lens, writing was seen as part of larger system in which the individual is connected to others culturally, socially, and institutionally.

DISCUSSION

These three instructors' courses demonstrate how writing was positioned among other activities in the classroom, focusing on how the tasks worked relationally to facilitate disciplinary thinking. Writing is not a neat and tidy activity; it is the embodiment of many social and cognitive streams of activities: reading, speaking, observing, acting, and writing. Each of these professors, all in the same College of Engineering, approached writing in their course differently, from writing as transcription to writing as ecology, where mediation occurs through multiple connected texts (Spinuzzi 2003). All three required students to practice principles of their discipline in simulated lifelike situations, emphasizing the discourse and skills needed to progress through the major. The chemical engineering professor assigned activities as if they were related, but assigned them separately, without any overt connection in instruction or pedagogy. In the interview, he expressed concern about student writing ability, but practically writing served as a conduit for information. It occurred *after* the activities of engineering. That is, writing was assigned so that students could learn, much like the Shannon-Weaver Communication Model (Shannon and Weaver 1948), to produce a message without noise or interference. A message is generated and transmitted, so that someone can respond, with communication occurring much like a tennis volley. Its success and usefulness depends on its receipt and clarity. Using a "Fog Index," this professor positioned writing as an object, something to be controlled through succinct prose, and evaluated with a mathematical formula. Student success appeared to be based on the correctness of engineering material and a system of writing that is prescriptive and confined to words and sentences on a page. The writing assignments did not ask students to research their own writing or even that of others, and thus, students quite possibly missed the opportunity of developing "a stance of inquiry" toward writing (Downs and Wardle 2007, 577); a letter grade merely indicated how well the student had internalized the prescribed formulas.

Unlike this professor who acted as a gatekeeper of knowledge, the materials science professor was a mentor of knowledge, one who enculturated students into some of the profession's practices. His course was designed as a laboratory of experiential learning in which students were both exposed to engineering discourse and taught to use such discourse in multiple activities over time with immediate one-on-one instruction through feedback. Instead of focusing on "basic" writing skills, the instructor emphasized learning to write in engineering as highly specialized and contextualized. In addition, assignments were viewed as opportunities for students to understand theory after reading and discussing it in class. Through performing experiments, students were asked to enact and experience the theoretical abstractions about which they previously had limited knowledge. Moreover, through writing students could enhance their understanding and memory of engineering concepts while at the same time learn the appropriate disciplinary conventions of genres. This professor consciously moved students toward the workplace by requiring them to practice engineering principles and apply them in their assignments. For example, a formal report with a literature review, he commented, would be something an engineer would write if he or she were publishing, which necessarily involves more theory. On the other hand, there are occasions when theory is not explicitly needed in a report, for example, if a supervisor asked an employee to rate certain tests, the employee would report back those results with no additional text other than to explicate them and make recommendations. Certainly, theory is necessary in both cases to complete the work, but the two uses of the report dictate different kinds of vital information. Regular meetings with every student each week to provide feedback and guidance on her or his engineering and writing enabled this professor to discuss how material is important and in what forms. Such practice, he observed, led students to develop a deeper understanding of what it means to be an engineer.

Finally, the professor in electrical engineering was similar to that of materials science in that he shared a comparable philosophy about learning: experience and practice, coupled with feedback, can move student thinking about engineering forward. Like the materials science professor, he integrated workplace documents into the course. Unlike the materials science professor he did not provide direct instruction, but instead provided models of professional writing from which students could glean information about the report genre.

The notebook, which he emphasized in the course, was used to keep a record of one's work. It reports contextualized thinking and problem solving, so if one needs to return to the creative process, information can be found that otherwise might have been lost. It also serves as a legal document for patenting. Interestingly, the notebook also served a much different purpose, as students were asked to refer to it as they wrote their final reports. In this sense, the notebook acted as a source for the creation of another written text, what Spinuzzi explains is an ecology, consisting of "assemblages" of genres that are used to "get work done: how [people] consistently marshal information resource and support tacit knowledge, how they search for information, and what they look for" (2003, 110). While Spinuzzi draws on research in the workplace to understand this dynamic interplay of genres, the electrical engineering professor intuitively asked students to immerse themselves in one genre in order to construct a second one. In this instance, the practices and strategies served to enculturate students into the specialized evidence-based discourse of engineering. And, writing in specific genres possibly functioned as portals to transfer disciplinary knowledge (see Nowacek 2011, for her "agents of integration" approach in which multidisciplinary contexts and contents provide opportunities for the transfer of knowledge and practice in writing).

CONCLUSION

We started this chapter asking about the goals and contexts for which professors assigned writing. Each professor articulated unique goals; underlying all of them was the tacit idea that students use writing as a means to identify with the discipline. None of the professors taught writing strategies such as the rhetorical "moves" made in research paper introductions; however, each addressed writing as a disciplinary pursuit with specialized terminologies, conventions, and genres (Hyland 2004). This echoes David Russell and Patricia Harms's (2010) research that found "genres. . . . [give] students the sense of the field, and a sense of themselves as engineers" (239). Professors in our study, like Russell and Harms's students, saw the strong link between writing and engineering and assigned various genres in their courses. But knowing there is a link does not necessarily translate into teaching the relationship in ways that facilitate students' enculturation into the field. It is not clear, for instance, whether genre knowledge acquisition and awareness enabled students to write more effectively in or across engineering contexts. Such inquiry was beyond the scope of this study.

Writing and engineering are often portrayed as being at odds. Our research demonstrates, however, that engineering professors (at least the three in our study) are conscientious about the writing their students complete in class. In our interviews, we found each of them had specific reasons for including writing in their courses: writing was a means to learn, memorize, and communicate and to facilitate generating other texts. All understood the importance of writing in the classroom as preparation for writing in the workplace. As such, the professors perceived the situated writing activities they assigned as communication instruments, disciplinary markers, and/or carriers of disciplinary practices. They claimed they used writing as a learning tool, reflecting a model that emphasizes disciplinary principles and contexts (e.g., Downs and Wardle's 2007 "Teaching about Writing").

Each professor indicated that they wanted students to think about writing and encouraged them to see how it was integral to engineering practice. Optimally writing was seen as a carrier of disciplinary practices, integrated into engineering practices, where doing engineering is also doing writing. They are seen as potentially constitutive. Such emphasis on production is crucial, as students participate in the creation of discipline-specific knowledge, rather than learning broadly about writing. The curricular designs in this study were somewhat more limited in scope than Rebecca Nowacek's transfer-as-recontextualization model (2011) of writing instruction, which offers a means for students to draw connections between content and genre by "actively working to *perceive* as well as convey *effectively to others* connections between previously distinct contexts" (38, emphasis original). These professors focused on language, content, and genre in their writing assignments; overall, little attention was paid to reflection—where students engage before, during, and after their writing process—to note "how they might apply the writing knowledge learned in the course to other writing [in this case, engineering] situations" (Yancey, Robertson, and Taczak 2014, 58).

Research suggests that students learn best in context, in the situations for which the presented information and material is relevant or becomes so over time (Artemeva 2009). Students are apprentices who, under the guidance of a master practitioner, can begin to perform accordingly and as a result socialize themselves into a discipline (Rogoff 1990), shaping it through the culture, history, and knowledge they bring to it as well. And, writing—like engineering—is complex, comprised of theories, practices, and dispositions. Teaching the relationships between the two can better facilitate students' socialization into their fields so that they not only know *about* writing, but *through* writing can create effective messages that resonate with others in their discipline.

NOTE

Mathison's data collection for this chapter was supported by a University Research Fellowship at the University of Utah.

3
TEACHING (EACH OTHER) (ABOUT) WRITING

Doug Downs

Engineers' relationship to writing is varied, as shown in the previous chapter. On a continuum, some see writing as a tool for putting thought into word, whereas others may position writing in the classroom as a tool for learning about the discipline. Yet, all three engineers demonstrated a lack of explicit connection between writing and its epistemic role in the discipline; there may not be a field that demonstrates a wider gap between what its participants *think* about writing and what they *do* with it. In this chapter, I argue that engineers *typically* conceive of writing as the grammatically correct, stylistically neutral, objective recording of factual information itself learned either through primary research or through other established sources (Winsor 1990a; Winsor 1990b; Winsor 1992; Winsor 1994; Winsor 1996). Because for engineers, writing is not typically understood as epistemic (generating knowledge) but is instead generally scribal—its function limited to the recording and transmission of knowledge—their concerns with writing tend to be largely formal: what form must the writing take to most clearly *transmit* the information efficiently and accurately? Yet, as Winsor has extensively demonstrated—when, for example, she tracked (1990b) a mechanical engineer's process of commenting on engineering documents and preparing a technical paper for presentation at a conference—engineering writing is *decidedly* epistemic and interpretive, and its success hinges on far more important matters than grammatical correctness and lack of typos. Thus, we see a significant gap between the actual nature of engineering writing (including how engineers actually accomplish it, and how they learn it) and engineers' typical cultural *conceptions of* the nature of writing (including how they think it ought to be done, evaluated, and taught).

One feature, then, of disciplinary writing instruction as cross-cultural interaction is how a discipline's cultural conceptions of writing impact the teaching of writing in that discipline. Our program had different kinds of conceptual teaching to do in different engineering disciplines, but the common denominator was that *each* discipline required extensive conceptual work to help the engineering and writing cultures understand what each other were assuming about the nature of writing and its teaching. (And to help the engineering culture understand that, and why, research on writing favors the writing culture's conceptions of writing.) In the Department of Civil Engineering, where I did most of my research and teaching, we had fewer conceptual hurdles than in most of the other engineering disciplines because of the greater amount of writing, and the more widely varied audiences for it, civil engineers tend to do as compared with most other engineering disciplines. (Mechanical engineering seems to peg the other end of the spectrum.) Even so, conceptual work constituted easily the largest percentage of my instructional efforts, not only in interacting with students, but in communicating with engineering faculty, in curriculum design, and in evaluating writing.

In this chapter, I explore differences in cultural conceptions of writing between engineers and writing instructors (with a particular emphasis on the role of metaphor in structuring these conceptions) and consider the role of those conceptions in cross-cultural, disciplinary writing instruction. I'll focus particularly on the implications of conceptions of writing for

- establishing writer instructors' expertise: what is there to know about writing and writing instruction?;
- determining instructional emphasis: what counts as writing, and what should writing instruction focus on?;
- achieving the desired instructional emphasis: what counts as writing instruction, and how can that instruction play out?;
- assessing writing: how should writing be evaluated, graded, and responded to?

In each of these areas, all of which I take up in detail in the second half of this chapter, differing conceptions of writing yield radically different answers, so that much of the intellectual work of disciplinary writing instruction lies in identifying and merging or bridging these cultural conceptions.

CONTEXT

Civil engineering (CE) is an interesting site for communication instruction, not only because it is among the most writing-intensive engineering disciplines, but also because civil engineers have some of the most widely varying audiences or stakeholders in the domain of engineering, and, as the name implies, a frequently more *public* and *municipal* nature than most other engineering disciplines. These qualities directly impact the teaching experience of communication consultants in CE, so it is worth unpacking them in greater detail before I analyze the experience.

The CE department lists its research foci as environmental, geotechnical, and construction materials, and structural, transportation, and water resources. Civil engineers work on transportation and water infrastructure that interacts closely with land and ecology, and in particular they design large, fixed structures such as roads, buildings, water treatment facilities, wastewater treatment facilities, dams, and pipelines. As such, CE tends to be about distribution of publicly regulated goods (such as water, gas, electricity), facilitating transport (roads, traffic systems, parking), and designing structures to national codes that require state inspection and approval. Therefore, civil engineers regularly work with the largest and most infrastructural federal agencies (and their state and local counterparts), including the Department of Transportation, Department of Energy, Environmental Protection Agency, and Federal Aviation Administration. And because few civil engineering projects are small scale (or entirely private), they usually require massive, typically tax-based, funding. By nature, then, CE communication has *a lot* of stakeholders. Civil engineers communicate, of course, with other professional engineers (both in and out of discipline) and with researching engineers. But they communicate almost as extensively with government agencies and inspectors, local government officials, major private and public funding agencies, politicians, public special interest groups, and citizens and taxpayers impacted by and benefiting from their work. Along with such a wide range of stakeholders, the stakes of a civil engineer's work themselves tend to be very high and very public. In being tasked with designing environmentally safe solutions to problems such as clean water and safe transportation, civil engineers routinely make design decisions that hold the lives of hundreds of thousands and the well-being of entire ecosystems in their hands, and upon which hinge the disposition of hundreds of millions of dollars.

To accomplish, say, the design and construction of a new treatment plant to bring a watershed into a city's drinking water system requires civil engineers to do a wide range of writing (and presentations) in three main phases. First, engineers "specify" (write specifications for) the project, which means carefully defining the problem(s) to be solved. (This happens at one level in writing the request for proposals for the project, and at a finer-grained level once the engineers awarded the contract actually begin their engineering documents.) Second, engineering firms bid for a contract for the project, requiring statements of qualifications and various proposals. Third, the contracted firm creates the actual engineering documents, which establish the plan to be followed at every stage of construction. The engineering documents combine design rationale and explanations, final specifications, and drawings. Documents are prepared in stages such as preliminary, 50 percent, 75 percent, and final, with increasing degrees of detail. (In the same way that wise writers do not copyedit until fine-detail work will not simply be erased by a section deletion, civil engineers work from large design issues to small ones, seeking client review and approval at each level.) While the bulk of engineering documents are drawings, every significant design decision by the engineers requires documentation. (For example, will the treatment plant purify water with an ion-exchange system or a direct-filtration system?) Major decisions are usually based on separate feasibility studies specifically conducted to let the engineers choose carefully among a number of options.

Such extensive writing written for so wide a range of stakeholders characteristic of CE raises two points of interest for communication instruction. First, civil engineering students must encounter a wide range of genres—*far* wider than the typical humanities major, for example. Along with standard writing-to-learn exercises such as responses to readings and brief essays on concepts and principles, our students practiced

- qualifications documents, including resumes and statements of qualifications in response to requests for qualifications, written to potential clients;
- correspondence, including intrateam and interteam memos detailing progress and decisions on divided-labor projects, written to fellow students, project supervisors, and teachers; and letters to clients transmitting documents and scheduling other interaction;
- proposals, in response to requests for proposals or expressions of interest (including both to bid projects and to request funding)—may include definitions of problem and scope, technical proposals, price proposals—written to potential clients;
- feasibility studies and reports on major design choices (for example, traffic management plans), written to project supervisors and clients (sometimes documents of public record);
- engineering reports on specific engineering problems encountered on the project, written to project management and sometimes for wider publication in the profession;

- field reports—such as site reports, surveyor's reports, and progress reports—written both to other team members and supervisors and for documentation (these will usually be documents of public record);
- specifications—transforming client objectives and guidelines, existing codes, and design-team parameters into specific requirements for materials and procedures during construction—written to other team members, project contractors, and the client;
- engineering documents at various stages of design, usually with letters of transmittal and reports that offer total project background and progress-to-date, in order to contextualize the work completed and work remaining—written to clients and other project contractors.

This list demonstrates not only the variation but the *volume* of writing that we helped prepare engineering students to encounter in their everyday work. (The text alone of engineering documents on projects of even relatively limited size can exceed 100 pages.) Documentation needs are tremendous: what did the writers decide, why, and what did they do to reach and test that reasoning?

The second point of interest regarding the extent and nature of writing in civil engineering is one of its results: expert civil engineers are expert writers who are highly aware of the need for engineering students to be strong writers and communicators. By some estimates, writing is 70 percent of a professional engineer's job (National Commission on Writing 2004), which civil engineering faculty, many of whom are also professional engineers, are perfectly aware of. What this meant for us teaching communication in CE was that, unlike in some other engineering departments, no one questioned the value of communication instruction or even contested the time spent on it—it was instead considered an absolute necessity, and the department was highly invested in it. Alone among engineering departments, CE actually had its own Technical Communication course. Because of the project-management aspects of writing in CE, professors also readily understood writing as a collaborative activity, not a private one, and therefore understood that an aspect of what students needed to learn about writing and presenting was how to do it *together*. Understandings such as these were not just the envy of writing instructors in other engineering disciplines, but would be the envy of instructors across the university.

What we writing consultants encountered resistance to, then, was not the *need* for instruction. It was easy to recognize the CE faculty as writing experts who had interesting and important things to say about how writing is the lifeblood of CE and were eager to build writing instruction into every level of the major. In CE, I worked with three courses: Introduction to Civil Engineering, Technical Communication, and Professional Practice and Design. In all courses, my role as the writing consultant was to collaborate on curriculum design and to provide occasional lecture-based instruction, occasional small-group coaching, evaluation of and response to all writing, and one-on-one consulting with students in a writing-center setting. The introductory course was designed to orient new majors to the discipline, with a particular focus on ethics and on the range of specializations within the field. Students wrote responses to readings and gave group presentations on material. The technical communication course was designed for direct, extensive instruction in writing and speaking, and I will say more about the writing instruction that actually happened there later in this chapter. The Professional Practice and Design course was a capstone professionalization course in which roughly thirty class members collaborated on a major design project contributing to actual construction (for example, the design of a low-impact/sustainable subdivision, a campus parking ramp, or a traffic management plan around a campus/hospital light-rail stop), producing a statement of qualifications, proposal and feasibility report, and final report with 75 percent engineering documents. Each major document was accompanied by a whole-class presentation to clients running the actual projects.

Unlike some other ventures in the College of Engineering, then, our foray into CE presented no difficulties in terms of cooperation and even eagerness to incorporate communication instruction. Rather, the challenges we encountered were in thinking together with faculty about *how* that instruction could and should be offered. As accomplished writers as CE faculty tended to be, they just did not think much about it explicitly or have an especially good language for translating their *doing* of writing into the *teaching and learning* of it. This observed disjunct between practical ability and (as I'll describe along the way) instructional sense is what initially led me to explore conceptions of writing as both a source of and a solution to tension between practice and instruction.

CONCEPTIONS OF WRITING

One of the most foundational principles of teaching is: How you understand a thing will shape how you choose to teach (about) it. In writing instruction, we know this both instinctually and historically. When writing is understood as primarily inscription of existing knowledge (recording thought) whose main challenges are stylistic (one of the more dominant understandings of writing between the Enlightenment and the 1960s), instruction will emphasize matters of form and style (outlines, syntax, diction, orthography, correctness) while attending far less to the whole range of rhetorical concerns, including rhetorical situation, invention, and iterative development of ideas through drafting. (See, for example, Downs 2013.)

A central question for writing instructors, therefore, is how they conceive of writing—how do they understand its nature, how do they expect it to happen and to go, what makes writing "good," and what are their dispositions toward the activity? In this section of the chapter, I will present first a fairly typical (and thus I hope noncontroversial) set of conceptions of writing based on an amalgamation of current composition and rhetorical theory, to establish an agreeable baseline of how many if not most experienced composition instructors might understand writing. I will then contrast that with an account of the conceptions of writing I encountered in CE, linked to other research demonstrating similar conceptions field-wide.

Rhetorical Conceptions of Writing

As described by Paul Prior (2004), writing is the activity of composing (inventing and designing) and inscribing (arranging and recording in some medium) a language-using text. Writing processes, which are specific to a given circle of writers and readers in a particular writing situation, involve, in Daniel Perrin and Marc Wildi's (2010) analysis, phases of goal setting, planning, formulation (drafting), and controlling (feedback/revision loop). These phases typically overlap, producing the composing effect Ann Berthoff (1990) calls "allatonceness." Because as Dennis Baron (1999) argues, inscription requires use of tools (and composition can be aided by them), writing is inevitably technology based.

Writing is a form of symbolic human interaction (interaction via linguistic symbols tunneled via inscribed orthographic symbols). Because such interaction is inherently rhetorical, writing is, furthermore, a rhetorical activity; to be rhetorical is to be *situated, motivated, contingent, material/embodied,* and *epistemic.* Unpacking these five terms yields a reasonable description of "the nature of writing."

First, writing is *situated* in specific contexts and reasons-for-being that are unique to each act of writing. Writing is, as demonstrated by Keith Grant-Davie (1997), exigence-driven: people write only when some need exists for that writing. (Including writing produced to fulfill a private, psychological need for self-expression.) The meaning, conventions, form, and success of texts constitute and are constituted by the communities and systems whose work the texts mediate. As explained in chapter 2, this social dimension of exigence can be understood in terms of *activity*, where writing is a tool used by groups of people to help accomplish some task (see, e.g., Russell 1995; Russell 2010); in terms of *genre*, where recurrent forms of writing are dynamically shaped by the situation (see, e.g., C. Miller 1984, 1994a; Bazerman 2004); and in terms of *discourse*, where the language shapes the communities using texts, and vice versa, in what James Gee (1989, 1999) calls Discourses, "saying (writing)-doing-being-valuing-believing combinations" (1989, 6). To say, then, that writing is situated is to say that a text cannot be understood independently of the context in which rhetors produce and read it; what a text means and how it is composed and produced inevitably depend on the context and circumstances of its composition, production, and reception.

Second, if writing is situation- and exigence-driven, it follows that it is also *motivated*: rhetors have motives for writing what they write how they write it and for reading texts as they do. The production of texts and the meanings rhetors take from them, then, unavoidably depend on motivations. Because there is no unmotivated writing, there is no objective, neutral, unbiased, impartial writing or text. As the saying goes, everybody's got an angle. No angle, no writing.

Third, given that writing is situated and motivated, its nature and processes are thus *contingent* on the rhetorical situation and the motives embedded in that situation. The answer to most questions about how a text should be is, "It depends." There are few universal rules for how to write; instead, the common ground across writing situations—the guidelines and principles writers begin with and return to—is *questions*. What is your exigence? What must the text do? How will it be used? What are your readers' expectations and values for what will make the text "good"? How do these arise from the activity the text is meant to mediate? What genre does the situation suggest or require, what are its typical elements, and what modifications does this situation suggest to that pattern? What conventions—in diction, design, style, and mechanics—should the text follow? Such questions are the most stable, "universal" aspect of writing.

Fourth, writing is *material* (as rhetoric itself is *embodied*). Because writing is in one sense a thinking activity, it is easy to ignore how it must play out on material objects, and how it results from material (not just ideational) labor. This aspect of writing's nature is most visible in its technological aspect. Writers are proficient with certain kinds of material practices and objects, and usually the wider a range of materials a writer can work with, the greater her or his ability as a writer. To write is to be able to manipulate the technologies and materials of writing.

Lastly, writing is *epistemic*: it generates new knowledge. In most corners of western culture, the default conception of writing is as the recording of *existing* knowledge: first we learn something, then we "write it down"; or, first we do research and reach findings and conclusions, and then we "write up" our research, the implication being that we learn nothing new during and by means of the act of writing itself, which is conceived purely as recording. This conception emphasizes the *inscribing* aspect of writing and elides its *composing* aspect. But in both theory and practice, writing is generative, not simply scribal. In theory, reality constitutes and is constituted by language: our understanding of our world depends in part on the language we use to describe it, so that our expressive choices generate (understandings of) reality. In practice, writers usually find that the act of "writing down" ideas usually creates new ideas—inscription actually tends to be generative. It is because writing is epistemic that texts naturally develop iteratively through revision, each new version incorporating ideas that arose during the writing and reading of the preceding version(s). And in this conception of writing, revision is therefore natural, inevitable, expected, and *part of the plan*. If writing were not epistemic, then revision might be understood rather as error correction, avoidable by "getting it right the first time."

To summarize, then: In a rhetorical conception of writing, writing is a complex, situated activity in which a writer, in collaboration with other writers and readers, simultaneously and iteratively manages development of ideas, choice of most fitting language related to those ideas, design of the text to be inscribed, and production of the text itself—all in response to and formation of a situation that creates particular constraints, conventions, and demands for the text, to the end of creating texts that readers will be able to use to help accomplish particular activities and which texts then, in turn, continue to shape future situations, activities, and texts. This description makes writing seem pretty complicated—far more complex than would be anticipated if our conception of writing involved only inscription and not composition.

Scribal Conceptions of Writing

The starkest contrast to a rhetorical conception of writing is one in which writing is the alphabetic recording of language—the basic, fundamental skill learned early in grade school, covering what Perrin and Wildi (2010) call "graphomotoric" functions and part of what Prior

(2004) includes as "inscription." This scribal conception of writing covers what is universal to *all* modern prose: word choice, sentence structure—including, prominently, punctuation—concordant with the writer's language of choice (English, in this case), paragraph structure with transitions for flow between ideas and through an argument, and—drawing from these elements—an overall understanding of and concern with writing as *form* essentially split from content. In this conception, "writing" is separable from *what is written*, and so "writing" is the product of a set of basic, scribal skills that are themselves ultimately masterable by the end of high school. (Writing, by these lights, is perfectible, simply by virtue of not breaking any rules.)

The scribal conception of writing accords with what I have elsewhere (Downs 2004) called a "hardware/software" model of writing—"writing" is a sort of hardware that can "run" any software—a container, if we use a container metaphor such as those described by George Lakoff and Mark Johnson (2003) and Philip Eubanks (2010), into which any ideas can be "poured" without influencing the shape of the container, just as software does not change the shape of the hardware that runs it. Simultaneously, this scribal conception of writing often relies on a conduit metaphor, imagining writing as a transparent and pure pipeline conducting ideas from source to recipient without coloring or contaminating them in any way, so that the reader receives precisely and only what was sent.

It is this hard separation of form and content that lets us imagine "writing" to be a masterable skill apart from what is written about, so that a "good writer" would look good independent of the subject of the writing. Nothing, of course, could be further from the truth—see Richard Haswell (1991) for an extended analysis of "regression" in written syntax that occurs when experienced writers are tasked with writing about unfamiliar subject matter. (Quality of syntax positively correlates with familiarity with subject.) But scribal conceptions of writing nevertheless hold separate written syntax and subject-matter knowledge in texts, a conviction endlessly reinscribed in high school education that leaves most students equating writing with syntactic correctness and obedience to formal rules (most notably academic documentation styles).

The historical roots of this scribal conception of writing are worth reviewing for the insight they give us on this form/content split. Historically, rhetoric with its five canons (invention, arrangement, style, memory, and delivery) encompassed the entire intellectual journey of "writing" (composing) an oral address (inscription was unusual back when Aristotle taxonomized the canons, hence the prominence of *memory* among them). Invention and arrangement in those times are both understood as *epistemic*, sources of knowledge, answering the question of how people come up with what to say—until Peter Ramus revolutionizes education in the mid-1500s by radically restructuring the academy. First, he essentially invents academic disciplines as we know them today, anticipating the division of natural philosophy into the sciences, and explains how to teach the disciplines through principles of dialectic—taxonomy and deductive generalization. Second, with content knowledge now the terrain of disciplines, invention and arrangement become matters of *dialectic*, not rhetoric. Rhetoric, Ramus decides, properly has no epistemic function; it deals only with the *expression* of knowledge garnered in the disciplines. Henceforth, "rhetoric" would include only three canons: style, memory, and delivery. This division of content and form pleases Enlightenment empiricists such as Francis Bacon, John Locke, René Descartes, David Hume, all of for whom language is a stumbling block to knowledge rather than a path to it. For these newly "scientific" thinkers, truth exists independently of its expression, and the difficulty of "style" is in *staying out of the way*. Bacon and Locke, in particular, hunt long for the *clear* language that would not color the scientific truths of empirical observation when writers attempted to express them.

Thus, Rhetoric labors through the seventeenth and eighteenth centuries with three canons, still focusing heavily on oratory and elocution. But by the nineteenth century, with printing presses and cheap paper ascendant, writing, not oratory, is becoming the primary mode of learned discourse. Rhetorical instruction similarly shifts, driven, as documented by Russell (2002a), by three main academic genres: the research report that then marks intellectual labor in the German university model, the freshman theme that characterizes U.S. college writing instruction at the turn of the twentieth century, and the student research paper that caricatures the scholarly research report at the undergraduate level. Memory and delivery being at the time more suited to oratory than writing, those two canons depart writing instruction (and Rhetoric with them, traveling with oratory to departments of Speech or Communication). "Writing" thus is at that point reduced to one canon: style. (See Connors 1997, for a far richer description of this movement, and Maureen Goggin 1997, for a history of English departments and their subjects.) Whatever teachers said about writing, what got the most red ink were issues of syntactical felicity and formal correctness.

The reduction of writing from the development of new knowledge (five canons) to the accurate, clear inscription of existing knowledge (one canon) lets Edward Frank Allen write this in his 1938 *How to Write and Speak Effective English*: "Perhaps the greatest obstacle to the achievement of proficiency in writing is the inability of the untrained writer to recognize an error when he sees or makes one" (14). Allen is working from the radically narrowed, one-canon conception of writing left after the rise of hyperrational empiricism and ubiquitous writing had ravaged fuller rhetoric conceptions. This definition of writing, at a nearly instinctual level, as a matter of syntax and style, particularly in educational settings, remains ubiquitous—so much so that Bob Broad (2003) found that academic readers assessing student texts would misidentify content problems as syntax problems when there was no easy way to describe the content problem but it happened in the vicinity of an easily identified syntax problem. Haswell (1991) similarly concluded that teachers seem to believe that "[complexity of] syntax and logic account for little, while length, error, vocabulary, and concreteness account for nearly all" of what makes writing "good" (41).

The typical engineer's conceptions of writing are far closer to the scribal than rhetorical, privileging accuracy and correctness over writing as generative. For example, Winsor (1990a) found that though most engineering texts are actually based mostly on other texts, engineers tended to perceive invention in their writing as based mostly on data from primary research. Engineers tend to believe in the possibility of objectivity in writing, as Watanabe's (chapter 9) description demonstrates. Read and Mathison in chapter 4 demonstrate how Read encountered the form/content split in engineers' thinking about writing, as she describes a common model for grading writing in engineering courses: English "graders" were hired to grade "the writing," while engineering faculty graded its content—and "good writing" was taught as "clean, correct" writing. "Content equaled engineering and form equaled writing," and writing equaled "grammar and style," they explain. Even the not subtly clever original acronym of the Center name alluding to clarity reinscribes the holy Baconian grail of the style to which engineers aspire: a transparent medium for the transmission of truth discovered by other, nonlinguistic means.

I conclude this discussion of conceptions of writing by describing the inaugural edition of CE's upper-division Technical Communication course. We communication consultants were brought in on the course after it had already been designed, approved by the university, and scheduled for its first offering, and therefore we had limited input in curricular design. The instructor had developed his own workbook combining a self-written usage/grammar/punctuation handbook and exercises for students. The course was reminiscent of earlier scribal writing instruction, with an emphasis on grammar, spelling, and punctuation, a pedagogy no longer aligned with writing theory. The class had sixty students. In a sixteen-week semester it gave twelve main "writing" assignments, nine designed by the instructor and course TAs and three designed by me in my role of writing consultant. The nine assignments designed by the engineering professor were exclusively short proofreading and editing assignments—harking back to the Ramistic notion of communication and rhetoric as purely formal, not having to do with developing content, invention. Had I not been a consultant to the course, students would have composed *no* prose of their own; it would all have been workbook assignments in copyediting prewritten prose. Even with my intervention, students did no work in actual engineering genres (e.g., reports). Instruction was entirely lecture and exercise based—there were no workshops in the class save one linked to the one significant assignment (a documentation project) I designed. There was no postfeedback revision of assignments. The course, in short, was a tribute to scribal conceptions of writing.

By way of transition to the third and final section of the chapter, I want to return to this curious bifurcation in engineers' relation to writing. On the one hand, they *think* and talk and teach about writing in these scribal ways. On the other hand, as professional writers, they are *doing* writing rhetorically (iterative processes, epistemic invention, the works). The disjunct is tremendous. And it shows up in other research, too. D. Cunningham et al. (2010) demonstrate that engineers do evaluate writing primarily on content rather than on grammatical correctness and spelling (16); Nicole Amare and Charlotte Brammar (2005) explain the disjunct in some measure by showing that engineering professionals tend to read for content and organization, while instructors tend to read for style. What the findings all converge on is this point: engineers' conceptions of writing lead them to design writing instruction that is neither reflective of how writing works nor of how engineers work as writers, and the cross-cultural play of disciplinary writing instruction is therefore *an act of revising those conceptions*.

THE PLAY OF CONCEPTIONS IN CROSS-CULTURAL WRITING INSTRUCTION

As noted at the beginning of the chapter, I am focusing here on four areas in which our negotiations between scribal and rhetorical conceptions of writing occurred: writing consultant expertise, instructional emphasis, instructional activities, and assessing writing.

Consultant Expertise

The first negotiation with CE faculty in which conceptions of writing played an obvious role was, naturally, turf: building a WID program from the ground up requires figuring out a lot of logistics, including the inevitable "who is covering what?"—a subset of which is "who *can* cover what?" In a scribal conception of writing that splits form and content, the default starting point is ever "You teach the writing and we'll teach the engineering." Of course, a rhetorical conception of writing finds this impossible: there is simply no way to cleanly cleave syntax from subject. At stake here is a notion of *organicism*. Writing, in rhetorical conceptions, is an irreducibly complex, organic weave of many ecological elements and factors shaping each other simultaneously, in the same way that in a living organism nerve cells, muscle cells, and capillaries are tangled together in incredibly dense mattings—not neatly divisible into muscle, nerve, and circulatory "systems." Scribal conceptions of writing are practically born to resist this organic character, to deny that subject matter shapes syntax simultaneously with syntax constraining subject matter, exigence and context shaping both of those even as both of those bend the exigence. Scribal conceptions desire composition and inscription to be separate events in order to reduce the "allatonce" complexity of writing. But that complexity remains irreducible. The intricate interplay among invention, arrangement, inscription, and design has each constantly leading to and giving way to the others. What the engineers would have asked, had they had this language, would have been "let us teach invention and you teach copyediting." As writing consultants, we demurred; the inseparability of content and form meant that if we had no role in content, we had no role in form.

Yet the saving grace in such meetings of conceptions was, for me, the assurance of the engineers' own experience with writing. When we asked them to describe their own writing experiences, it became much easier to say, "Our research and your experiences both suggest that the way we are going to need to teach is by X, rather than by the Y you are assuming." ("Our research and your experiences both suggest that we need to take a different approach to enculturating students into disciplinary ways of thinking through writing.") These moments weren't *change* in either of our conceptions, but instead deflection: engineering faculty were willing to accept, provisionally, a trial of a counterintuitive approach based on the writing consultants' assertions of research-based expertise (as long as it did not openly conflict with the engineers' own writing experiences). This solution would do little, at first, to prevent the same scribal-based suggestions the following semester, but it let us bring our values to their house.

Setting up this shared conceptual "housekeeping" took negotiation and collaboration but worked quite well. Based on Haswell's (1991) research and theory on regression and how increasing subject-matter expertise tends to improve syntax, I argued that if we focused extensively on content, the syntax would come along mostly on its own, which did turn out to be the case. In early engineering documents in the Professional Practices course, for example, we saw uncomfortably high error rates as students concentrated primarily on solving engineering problems they didn't yet fully understand. But rather than assuming the students didn't know the rules they were breaking or were truly unable to proofread, I coached faculty that the syntax would improve as students better understood what they were trying to say to begin with, and if we simply stressed the responsibility of the writers to proofread. Later engineering documents were indeed far more fluently written and better edited, sufficiently so to be nonembarrassing when delivered to professional engineering firms. In fact, in no class that I entered did I find systematic or widespread syntax problems, which increased my confidence in reporting to engineering faculty that we should instead focus on students' ability to understand and predict what readers would need to know, so that they could then organize and design their documents to meet that need. The weaknesses were *rhetorical* rather than *grammatical*. This was most clearly the case in the capstone Professional Practice course, where the rhetorical complexity demanded by the design tasks of the course exposed students' propensity for writer-based prose. Students too frequently forgot that readers are not mind-melded with writers, and so in early engineering documents we often found instances of vague reference or missing data because students were assuming that their readers had seen everything the students had. There is nothing grammatically wrong with the sentence "We will bury the pipe under the roundabout," but if neither the pipe nor the roundabout has been referenced previously in the document, the use of definite articles to refer to those objects is guaranteed to miscue readers into wondering what they've missed. *What* pipe and roundabout?! The "grammatical" teaching here would be "don't use *the*!" The *rhetorical* teaching would be "stop and think about the *context* in which you're writing—what do and don't your readers know yet?" The students in such cases didn't need syntax or grammar lessons, but rhetorical lessons.

Following this logic on the inseparability of form (syntax) and content (subject), eventually engineering faculty and communication consultants shared *every* aspect of instruction—curriculum design, teaching, and responding to writing. The engineers were able to teach us sufficiently about what constituted good content to the extent that we could contribute to planning discussions and give students good advice directly; and we were able to read the resulting writing with engineering faculty and offer a language for describing and a diagnosis of the problems we all identified in students' writing.

Instructional Emphasis

Naturally, while we negotiated the problem of *who* would teach what, we were also working on the problem of *what* to teach. From the standpoint of conceptions of writing, two foundational questions were "what counts as *writing*?" and "how should we *model* writing?"

The first question isn't terribly complex. In scribal conceptions of writing, *inscription* counts as writing, and so it is possible to teach "writing" without ever asking students to compose a document. *Writing* in this sense elides genres and documents for a formal focus on sentences and the skills of editing and proofreading. This scribal answer to the conceptual question "what counts as writing?" was almost entirely responsible for the original design of the Technical Communication course. I wanted to understand "writing" more broadly, considering it both in terms of composed products (documents) and in terms of composing and inscription processes (the entire timeline of where writing comes from). One key to our collaboration, then, was sharing critiques of the course after its first offering. My theoretical concerns aligned well with student feedback reporting that Technical Communication had offered him or her no preparation for the writing assigned in his or her ensuing capstone Professional Practices course. Through trial and error and collaborative critique, which blended multiple data sources, faculty were able to build a more rhetorical, less scribal sense of what counts as "writing."

The second question—"how should we model writing?"—was more complex, because it wasn't the actual question. The actual question was, "what does improved writing instruction look like and how should it proceed?" But the answer to that depends on a question more like "what writing processes do we want to make visible during writing instruction?" In other words, what aspects of writing should enter the classroom? Based on rhetorical conceptions of writing, writing instruction would make *every* phase of the writing process, from invention through proofreading, visible in the writing classroom. One way to show and model inventional and revision aspects of writing is to conduct interactive, collaborative workshops of one form or another. The question of what to teach, however, leads straight to how class time will be used: lecture is not sufficient. But from a scribal conception of writing, what needs to be shown about writing—rules about form, grammar, and punctuation—can be covered with a PowerPoint and a website full of handouts.

We encountered an interesting disjunct, then, between the Technical Communication and Professional Practices courses, because what needed to be taught in the two courses was, to engineering faculty, clearly differentiated—"writing" versus "content." Because engineers tend to start from a lecture-based instructional model, the default for what aspects of writing to show was whatever can be lectured about (see Watanabe, chapter 9). Tech Comm, with its focus on grammar and copyediting, could "get away with" such lecturing. Professional Practices, however, could not. Engineering faculty, given their own professional writing experience and expertise, understood that students needed to be shown writing behaviors and knowledge beyond editing, proofreading, and grammar and that lecture might be insufficient to model this broader range of writing knowledge and practice. At first blush, there was simply no time in the course to work on so wide a range. This problem bled over into the next major element of instructional design in which conceptions of writing played a significant role.

Instructional Activities

The same conceptions that led to interesting and difficult discussions around *what* to teach yielded similar debates about *how to* teach it. Because the engineering faculty's scribal notions of writing carried over to writing instruction, the battle that writing consultants most consistently lost was for instructional space for drafting and revision, work shopping and reading response, conferencing—essentially, any instructional activity other than short lectures, handouts, and feedback memos.

The "solution" we began exploring in CE, driven by rhetorical conceptions of writing as a long process of iteratively developing ideas through interaction, was the mildly guerilla tactic of taking writing instruction out of the classroom altogether. Our final report from the first year of writing consulting in CE contained these lines to be read between:

> In [CE], we have found the . . . lecture-and-response model to be generally ineffective, allowing students and faculty to imagine communication as "skills" rather than integral, discursive knowledge and behavior, and relegating it to "basic" status. We believe the model sets consultants up for failure both in [poor] teaching outcomes and lowered ethos. We have begun shifting from the lecture format that typifies engineering

education to more experiential, interpersonal instruction that forgoes most lecture in favor of meetings with individuals or teams. We encourage the Center to critically evaluate the effectiveness of "talking-head" lectures by communication consultants in engineering courses and work with faculty to develop alternative methods of instruction which more closely resemble best-practices (team work, peer review, recursive drafting with direct consultant feedback) in writing and speech courses.

Our difficulty was that, as innovative as integrating communication instruction into class time was to begin with (through occasional lectures), there was simply no culture for "interrupting" the "normal" flow of class lecture and team-meeting time with reading workshops, collaborative document review, or writing time. *There was no culture for such approaches because no conception of writing that would intellectually or viscerally support such a culture had been developed.* To develop the culture, we would have needed to change the conceptions of writing linked to it.

Metaphors were key to such change—being able to put rhetorical conceptions of and assumptions about writing in terms recognizable to people working with different conceptions. As a result, using the available instructional mode of handouts, I developed a list of metaphors for writing that would liken my conceptions of writing to ubiquitous ideas in engineering. In a "Ways of Thinking about Writing" handout, I drew metaphors equating various principles of writing to various subdisciplines of civil engineering. The principle of *context* I likened to environmental engineering, because of the need to understand how one element (say, a piece of writing) would interact with its surroundings in order to know how to shape that element. *Assumptions and values* became geotechnics: the foundations and grounds underlying a structure. *Scope and pacing* equated to water: making sure the size of the pipe corresponded to the desired flow, "sizing your writing correctly for the size of the ideas it carries." The implementation of a container metaphor, while theoretically not ideal from a rhetorical perspective, gave me grounds to converse with students in terms they could actually visualize based on their own discourse.

What such teaching did, somewhat effectively though subtly, was help students begin to recognize that there was more to worry about with writing than simply style. In other words, if there was no good way to meet scribal conceptions of writing head-on, at least it was possible to feed students *additional* experiences, evidence, and conceptions that weren't easy to integrate into the scribal conceptions. These additions would serve to (1) make bridges with existing conceptions, (2) start a dissonance that could grow into awareness of shortcomings of existing conceptions, and (3) provide points of connection and the beginnings of a latticework for later experiences to be made sense of in some light other than scribal conceptions.

Writing Assessment

The final major instructional element shaped by our conceptions of writing was our assumptions about evaluation of and response to writing. All roads lead to evaluation of writing, of course, and each preceding element has had something in it that would lead to challenges for reading and responding to students' writing.

Probably the most obvious is the aforementioned "you grade the writing, we'll grade the engineering" expectation. As I argued already, this approach makes sense to scribal conceptions of writing but not to rhetorical conceptions. Because with even the most cursory examination, "writing" (syntax) is empirically inseparable from content (change the expression of an idea, change the idea expressed), this was an argument we communication consultants could readily make clear. In short order, the argument was not *whether* consultants should grade holistically, but *how*—what kinds of response systems we could create that would allow both knowledgeable engineering TAs and communication consultants to *collaboratively* evaluate student writing and offer wide-spectrum feedback on the whole range of issues that a rhetorical conception of writing would consider important. This collaboration—which eventually meant "team-reading" students' work in the same room, in real-time consultation—usually took the shape of questions for each other. I would often say, "I read this section as making a poor argument because I can't see how the writer moves to this conclusion—what would you say as someone who already knows what the writer is trying to say?" Often the response would be, "No, this is the way an engineer would make this argument, because here's what we already take as given that completes the argument." It was crucial in these reading sessions for consultants to listen and learn all we could (without being able to follow much of the math)—we were trying to learn the Discourse, as James Gee (1989) would say—the sound, look, thinking, and behavior of engineering arguments. In return, consultants could contribute increasingly accurate diagnoses of the source of writing problems. If together we readers concluded that a section of writing failed to make its point or be readable, I would usually be the one to be identify why and thus present alternatives.

That ability links directly to the other element of responding to writing that seemed conception driven: the nature of feedback to students.

Engineering faculty seemed unused to giving positive feedback on writing. Correction of student writing was typically understood as calling students' attention to the consequences of unclear writing, and in that context the feedback style that made the most sense was a relatively harsh comment wherever the writing "broke." Early in my consulting role, I accepted this cultural norm and let it shape my own feedback style; this led to many comments on student work that could most readily be summarized as "your lack of attention to detail in editing your writing makes you look lazy. If this were an actual document in an actual job, you'd be fired."

However, as engineering faculty began to value responding to a wider range of issues than just grammatical and correct syntax, and as I as a writing consultant was able to offer more nuanced explanations for what made a text more or less readable or a better or worse argument, we began to be able to offer students richer feedback that was positive when it could be and constructively critical as necessary—especially written with an eye toward revision or transfer to the next writing project.

IMPLICATIONS: BRIDGING AND MERGING

The obvious lesson to take from the experiences described above is that fostering a cross-cultural sharing of conceptions of writing is not a *prerequisite* for successful disciplinary writing instruction; rather, it *is* writing instruction. Instructors preparing to enter another discipline or culture to teach writing should understand that a significant portion of their teaching burden will be not procedural instruction but conceptual instruction: working on how students and faculty in the other discipline imagine and understand the nature of writing.

Ideally, that conceptual work will take the form of *merging*: disciplinary faculty and writing instructors will shape their conceptions of writing into a shared understanding that accounts for what each culture knows about writing. A merged conceptual set offers the best chance for students to hear a unified, focused message about writing, consistent both with what writing experts know through research on writing and with the language, attitudes, and instructions of their own discipline's faculty. It is important to note that "merging" is not *colonizing*: the idea is not to purely *transplant* the writing instructor's conceptions of writing in place of the discipline's. Rather, in this ideal, the writing instructor learns much from the discipline that shapes the instructor's own conceptions into a different set than they were before the cultural encounter, and the same is true in the other direction.

One path to such merging of conceptions involves, first, *bridging* them: trying to put one set of conceptions in terms of another or in terms at least accessible to another. This is where the powerful work of metaphors emerges, as metaphor allows a writing instructor to restructure her own conceptions in terms already familiar to the culture she is working with. (And, again, vice versa.) Metaphors create opportunities to *recognize* conceptions to begin with: the comparisons they make expose assumptions and serve to mark, with high visibility, embedded values and attitudes. Thus, a single metaphor can expose a conception, offer a window for analyzing it, *and* offer a means of translating a conception from one disciplinary language to another. It is this multivalent power that makes metaphors so effective in bridging conceptual schemes for writing.

In any given circumstance, bridging may be the best viable outcome of a meeting of cultural conceptions of writing. Merging is a high ideal, possibly unreachable in many circumstances, especially early on in a disciplinary teaching relationship or in a short-term collaborative arrangement. Bridging—as an important early, translational strategy toward merging—may in fact also be as far as it gets. That reality suggests that bridging cultural conceptions is even more useful than it might otherwise be.

Whatever the method by which writing instructors in other disciplinary cultures focus on conceptions, the point remains that such focus is crucial. These experiences in engineering demonstrate that without a focus not simply on *how to teach writing*, but in fact *what writing is*, we can expect months of talking past each other and the frustration and even failures that such cultural miscommunication leads to.

4
LOCATING COMMON GROUND FOR DIPLOMACY
Using Critical Thinking to Teach Writing

Sarah Read and Maureen A. Mathison

Implementing a Writing-in-the-Disciplines (WID) program in a college or department without precedent is challenging. First, a WID program that integrates a communication curriculum into an existing course structure necessitates substantive curricular and pedagogical change and innovation; and, second, while administrators and faculty in the discipline may be ostensibly in favor of initiating a WID program, this surface enthusiasm often masks deeply held disciplinary assumptions about the "how and why and who" of how the program should be implemented. Both of these challenges present what is essentially a problem of diplomacy. Diplomacy, Geoffrey Cowan and Amelia Arsenault (2008) explain, has multiple layers, with one—collaboration—especially beneficial when working across cultures. It can "breed social trust, foster norms and reciprocity, and create stores of goodwill" (23). In our collaboration with the Department of Chemical Engineering, diplomacy involved the art of strategically meeting and mitigating process-based resistance to what was otherwise generally endorsed change to incorporate writing more fully into the curriculum.

In this chapter, we recount and situate within the theory and literature of writing and rhetoric studies how a diplomatic process of finding common curricular ground resulted in a comprehensive curricular model and pedagogical approach for integrating writing instruction into the department curriculum. At the theoretical level Susan McLeod, a key figure in the Writing-Across-the-Curriculum (and Writing-Across-the-Disciplines) movement, advocates for integrating writing into disciplinary courses beyond "just a course offered by the English department" (1992, 6). At the practical level, however, successfully implementing a WID program is easier said—and written about—than done, in particular when the guiding paradigm necessitates substantive curricular change. Substantive change is necessarily time intensive, labor intensive, and political *within* any discipline, but the challenges multiply when disciplinary assumptions and values about what writing is and how to teach it have to be translated *through* distinctive disciplinary lenses. Similar to Downs's experience (as told in the previous chapter), the administration, the faculty, and to a limited degree the students all supported more communication instruction in their department. The challenge, however, was to locate the common ground with which both the engineering faculty and writing consultant could collaboratively move forward.

Immediately issues emerged that precipitated initial resistance to the project. First, engineering and writing represent two unique cultures, with distinct approaches regarding who has authority in the classroom. In writing, graduate students are generally the teacher of record for a course, responsible for delivering instruction, meeting with students, providing feedback to them, and evaluating their efforts. In engineering, TAs do not assume such duties; rather, they generally act as graders or assist in the lab. This division of labor posed a problem for the graduate writing consultant: as a writing expert, but also a student, she initially came up against faculty resistance to recommendations to integrate substantive process paradigm writing instruction into engineering courses.

Second, in-group (Engineering) and out-group (Writing), histories with and beliefs about teaching writing clashed. In general, the professors' existing assumptions about and practice of writing pedagogy were informed by their own experiences in graduate school and as engineering professionals and academics (Read 2011), which did not include writing consultants. In addition, their buy-in to discipline-specific writing pedagogy compromised the ethos of the writing consultant because it subjugated her identity as a "humanities person" to that of an "engineering person." How could one expertise be useful to another? In his account of the history of the humanities-engineering relationship, O. Allan Gianniny writes of a disciplinary disenfranchisement: "A century of activity [has left] the curricular role of the humanities . . . on the fringes of engineering education" (2004, 343). The lack of ethos of the writing consultant, therefore, was not personal, but a legacy of difference in relational histories and pedagogical practices between the humanities in engineering.

Finally, the considerable degree of change proposed by teaching writing based on best practices for WID caused some alarm. While professors in chemical engineering already assigned writing in their courses, they did not explicitly teach writing because conventional writing instruction—such as lectures, drafting rounds, workshops, and feedback—were difficult to integrate into the current class structures. As a result, writing assignments were required, but occluded for students' variations in the purpose and the audience of their writing, as well as their roles as writers. In addition, like students in Poe, Lerner, and Craig's study of MIT engineering students (2010, 25), nearly all courses required the same written assignments, most commonly a version of the department's standard for professional documentation: the formal report and the laboratory report. These reports functioned mainly to report student work to the professor, even in writing-intensive upper-division courses such as the senior lab capstone course, where students were positioned as proto-professionals writing to a supervisor.

Despite the fact that students understood the capstone course to be writing intensive because they had to write three to four formal reports and shorter form letter reports over the course of a year, senior students had a difficult time changing their role as a writer to that of a professional, or high-stakes, expert role because they were accustomed to the lower stakes of the "reporting" format they encountered in their lower-level courses. Students struggled with new expectations for professional performance in the capstone course when previous expectations were for accuracy and correctness and not rhetorical astuteness. Added to the students' challenge was faculty holding students to professional standards of correctness in evaluating their work without additional support for writing. With its emphasis on producing polished documents over helping students to navigate new rhetorical situations for writing, the paradigm and practice for teaching writing that were in place could be summed up as "teaching editing is teaching writing," or as a limited version of the largely abandoned prescriptive and product-centered model for teaching writing.

With its emphasis on correctness, the existing department curriculum structure set up resistance to both of the main models of writing pedagogical theory: (1) the rhetorical paradigm, because the assumption in engineering is that writing is a discrete skill acquired through a transmission model (such as a lecture and worksheets) independent of engineering knowledge, and, (2) the process paradigm, because of the perception that the process of writing (roughly speaking, that of prewriting, drafting, and revising) takes up too much additional course and instructor time. However, while their outmoded belief system about writing clashed with accepted models of writing pedagogy, the chemical engineering faculty intuitively understood the nature of WID as it was implemented in their courses—that the writing needed to be situated within engineering. This fact opened the door for diplomacy and, ultimately, for change in their approach to teaching writing.

Initially, teaching writing the way a writing and rhetoric studies instructor teaches it seemed impossible in a context that resists both the role and the pedagogical assumptions and paradigms that come with being associated with writing and rhetoric. Because any WID program, just like nearly any implemented writing pedagogy, is founded upon research-based practices for teaching writing as developed by scholars in writing and rhetoric *and* the practices of the discipline that is the pedagogical focus (chemical engineering, in this case), it was necessary to find common ground with faculty collaborators that drew from both disciplines. Clearly, mitigating resistance and engaging in successful collaboration with faculty to integrate writing instruction into their engineering courses required finding interdisciplinary common pedagogical ground, which could only be achieved through diplomacy.

FINDING COMMON GROUND IN CRITICAL THINKING

The idea that a curriculum based on writing and critical thinking could be the much-needed common ground for collaboration first emerged during a series of conversations with a professor of the senior project lab capstone course. He recognized that his students did not need elaborate instruction in lab report formatting or editing skills because most students already demonstrated relative proficiency in these areas by the end of the year. But while students did a satisfactory job in their formal technical reports of describing their lab experiments and showed an understanding of how to work the apparatus, they failed to manage the ambiguity and complexity inherent to experimental processes and data sets. Further, students showed a general weakness in interpreting what this complexity implied for solving problems, drawing conclusions, and making recommendations required for the assignment. As has been documented in studies of students in general, they did not consider multiple possibilities, or rival hypotheses regarding their results (Flower, Long, and Higgins 2000; Higgins, Mathison, and Flower 1992). The question became, then, how to get students to the stage where they are as comfortable interpreting data as they are describing it. Essentially, the problem came down to the students' weakness in critical thinking skills and in showing critical thinking through writing.

During our discussions, the professor of the senior capstone course brought to our attention a common site where this weakness in critical thinking shows up in student work: the conclusion section of a lab report, where students must tell the reader what their data means and what their recommendations are in terms of what should be done next.

He expected to see these sections increase in complexity as students wrote successive reports over the course of a full year of the capstone course. He expected to see them develop in the following way:

- First Report (descriptive): "Here's what we get; you (the reader) figure out what it means."
- Second Report (comparative): "Here's what we get and here's how it compares quantitatively to accepted results."
- Third Report (interpretive): "Here's what we got, here's how it compares to accepted results, therefore here's what we think it means." (e.g., Senra and Fogler 2014)

However, the capstone professor felt that he did not see this improvement in student writing over the course of the year. For him, this process of learning how to write a conclusion section embodied the intersection of the goals for teaching writing and the goals for teaching students to think critically as engineers. During these discussions, it became clear that "critical thinking" was the paradigm that we could adopt to align our goals with those of the engineering faculty in order to develop ways for teaching writing as an integrated part of teaching engineering.

Finding this common ground, however, did not initially guarantee success. As John McPeck notes, critical thinking is discipline specific: a process cannot be separated from its content (1990). "Just as there are different kinds of 'language games,' which stem from what Wittgenstein (2001) called different 'forms of life' (e.g. mathematics, morality, religion, art, etc.)," explains McPeck, "so there are different rules of predication, or 'reasoning,' if you will, which govern the different kinds of thought" (36). To become a critical thinker in a specific discipline, one must learn the language and symbol systems that guide reasoning and belief about the material. While engineering faculty agree that teaching critical thinking is important, and while understanding that writing is crucial to successful projects, engineers tend to emphasize the technical over the communicative aspects of practice, often believing that writing occurs after the design is completed (Burnett 1996).

This was a belief that the writing consultant ran into during a half-day seminar to introduce the pedagogical model described in this chapter. Faculty listened skeptically to my presentation and then asked questions that implied undisclosed resistance. For example, questions about teaching critical thinking via writing assignments were framed in terms of common fears about journaling-type writing being irrelevant in technical contexts, even though journaling had not even been mentioned in the presentation. It was not until later that the writing consultant realized that the faculty had thought that she was proposing that critical thinking could *only* be taught through writing, and not also through the acquisition and refinement of quantitative and computational skills. The error in this situation was her complicity with her assigned and assumed nontechnical ethos in which she behaved as if she had no domain over the technical side of the curriculum and, therefore, unthinkingly, did not acknowledge it. At times it felt as if the writing consultants and engineering faculty were so entrenched in disciplinary roles and behaviors that attempts at communication across the great divide felt futile—moments which ultimately reinforced the call to diplomacy for the sake of a worthy cause, the "third culture" discussed in chapter 1.

DIPLOMACY IN PRACTICE: ESTABLISHING PRECEDENCE AND BUILDING ETHOS

Diplomacy required both disciplines to step out of their comfort zones. The majority of the faculty with whom the writing consultant collaborated felt they had to relinquish some control of their classrooms for a pedagogy grounded in another discipline, even though the guiding philosophy of the program was that engineering faculty shouldn't have to *insert* new material into an already crowded syllabus, but instead collaborate to develop material that *integrates* the teaching of engineering, critical thinking, and writing. Similarly, the writing consultant had to stretch beyond her own disciplinary boundary to extend her expertise about engineering communication. To get started, the writing consultant used a summer grant to research and construct an annotated bibliography, *Integrating Writing, Critical Thinking, and Engineering Curricula*, in order to establish precedence for the work we were undertaking (in the interviews conducted for chapter 10, a professor mentioned this as a resource). The majority of the sources were drawn from highly regarded educational journals and conference proceedings in engineering. This served both rhetorical and pragmatic purposes: (1) to establish that there is precedence for engineers integrating writing into engineering curricula; (2) to demonstrate that a diversity of methods and approaches have been tried with experiences of failure and success as varied as the approaches; (3) to provide a resource of assignments and curriculum formats for the purpose of curriculum planning for any given course and, very important; (4) to build an ethos as an "expert" in writing pedagogy as it specifically applied to the engineering context, bridging the divide.

Another source for establishing common ground around teaching writing via the lens of critical thinking was the department's existing documents on undergraduate teaching and learning objectives.

Fortunately, the department teaching objectives already clearly articulated critical thinking as integral to the curriculum. As declared in its strategic plan: "Students will be able to *analyze* problems, *design* experiments, obtain solutions, *evaluate* information, and *communicate* results both *individually* and as part of a *team* [our emphasis]."

Emphasizing existing department goals as described in the strategic plan helped reduce resistance by showing how teaching critical thinking through writing in their extant curriculum could also meet ABET 2000 outcome objectives, which include critical thinking and communication. The goals of a critical thinking through writing curriculum allowed students to

1. learn how to think more critically about the engineering content of the course and spend more time with the material;
2. practice discipline-specific writing skills, including genres and jargon;
3. learn that the process of writing and the process of thinking are linked, that is, that the process of writing is not just the task of "reporting" on thinking that has already been completed, but can push problem solving and conceptual understanding to higher levels.

The department learning objectives supported the pedagogical aims of the WID consultants to integrate writing into the curriculum by threading it throughout the goals, which were already heavily focused on aspects of critical thinking (e.g., analyze and evaluate). This connection between critical thinking and writing is not unprecedented—many engineers recognize the connection between critical thinking, writing, and their discipline. For example, critical thinking is central to many of the objectives of an engineering curriculum, Patricia Ralston and Cathy Bays explain, which they argue can be taught through writing in engineering (2010). In addition, "Good writing," Pradeep Agrawal argues in his article "What is Good Writing?," has two defining qualities: "It must be easy to read and it must demonstrate critical thinking." Further, Agrawal recognizes that a writing consultant can play a "pivotal role" in "present[ing] critical thinking to students as the foundation of effective communication" (1997, 6).

As a result of the development of an annotated bibliography and a resource packet about teaching writing in engineering and further discussions with the professor of the senior capstone course, and other faculty, we began to design a curricular strategy to teach writing through the development of critical thinking skills (and vice versa) within the existing curriculum structure of the department across all four undergraduate years. This process required developing a model for integrating writing instruction that could accommodate and account for the existing curricular goals and structure of the department. We employed a "backward" design (Wiggins and Tighe 2005) process in which instructors first decide what information they want students to learn and then work backward through the curriculum to design a course that guides students to the end point. We applied backward design to the entire chemical engineering curriculum, beginning with the capstone course. By learning what knowledge and communication skills engineers were expected to exhibit as professionals, we could then begin to design a curriculum that would support student learning at every level.

The existing curriculum presented both obstacles and opportunities for this work. On the one hand, working with a prescribed curriculum provided the opportunity to understand how we could increase complexity in thinking and writing at various stages so that by the time seniors conducted their research experiments, they could interpret and communicate their findings to an interested engineering community. On the other hand, improving critical thinking and writing early in the curriculum were mostly out of our hands—the majority of courses in which students enrolled during the first two years were primarily large, lecture-based math and science courses outside of the department. In addition, the pace of the few courses students took within the department was maximized to cover a large amount of content, making it difficult to integrate the teaching and learning of writing. At the upper-division level, classes also remained largely quantitative and lecture based, although the practice-based capstone course provided opportunity for integrating a more robust writing pedagogy.

THE CURRICULAR MODEL: WRITING-TO-LEARN BEFORE WRITING-TO-COMMUNICATE

The model codesigned with faculty built a four-year curriculum that developed student critical thinking and writing skills beginning at the freshman level and continuing through the senior capstone course. The purposes for students' writing, however, would change over the course of the curriculum as the nature of their learning about engineering changed as well. The core philosophy of the curriculum was developmental: in general, students would be asked to write to support learning in the lower division courses that focused mainly on learning engineering content before they were asked to write to communicate their knowledge of engineering as proto-professionals in the practice-based senior-level capstone course.

The theoretical foundation of this four-year model was based on the classic distinction that writing studies scholars have long made between expressive and transactional writing (Britton et al. 1975; Martin et al. [1976] 1994). These two types of writing differ in the two dimensions that characterize variation among all writing: audience (*whom* is the writing for?) and function (*what* is the writing for?). While the function of transactional writing is to transfer information from the author to the audience, the function of expressive writing is to support learning free from standards of correctness. In transactional writing, the author assumes the role of expert and is held by the audience to a high standard of accuracy. On the other hand, expressive writing, with the audience as self or teacher-as-mentor, does not hold the author to the same level of expertise and exploration, and experimentation is the goal. Given these differences in audience and function, students learning new conceptual material will be able to write about it expressively before they are able to write successfully about the new material transactionally. Within this line of thinking, teachers are encouraged to provide assignments that require a variety of audiences and purposes for writing and to encourage students to consider writing as useful for thought and learning processes beyond the reporting of knowledge and evaluation.

Within WID, expressive and transactional writing can productively be thought of in terms of assignments that promote writing-to-learn (expressive writing) or writing-to-communicate (transactional writing). Writing-to-learn assignments encourage writing by prompting students to reflect, question, and process new concepts that they have learned from reading a text or listening to a lecture. The goal of writing-to-learn exercises is not to evaluate students' understanding of a new concept, but to encourage the students to test their own understanding of the topic, to identify areas of poor or misunderstanding, and to encourage students to think individually about a concept and relate it to their own context. Although some assignments may still be graded on effort and/or completeness, in general students must understand the assignment to be "low stakes"; that is, the assignment will have little impact on their grade, and they will not receive extensive criticism from the instructor. The objective of such an assignment is to aid, not evaluate, learning.

Writing-to-communicate assignments, on the other hand, convey information and are generally longer and more complex; they conform to particular conventions. They are transactional documents, the kind of document that "gets things done" in either a scholastic or professional (or even a personal) context. Writing-to-communicate assignments are expected to be more sophisticated and demonstrate expertise rather than just engagement with concepts. In the department, existing writing-to-communicate assignments included letter reports, formal technical lab reports, and research proposals. Students understood they would be evaluated on the correctness of the content of these reports, as well as on their ability to fulfill genre expectations, including presentation of the content. The audience in high-stakes writing is evaluative and has high expectations, whether the named audience is a professor, as is the case in a classroom, or a boss or manager, as is the case in a professional setting (sometimes in the classroom the professor plays the role of "supervisor").

At the level of implementation, a pedagogical model that is based on the idea that students write expressively about new concepts before they can successfully write about them transactionally implies that writing assignments in lower-level courses are primarily, though not exclusively, for the purpose of writing-to-learn, while in upper-level courses students begin to learn and practice the discipline-specific and professional forms of writing-to-communicate. It is also important that students learn early on the purposes of both types of writing and learn to be comfortable using expressive writing as preparation for transactional writing. This approach to developing writing assignments made sense to engineering faculty because they understood that students needed to understand conceptual material before they could engage in higher-order critical thinking to identify problems, analyze information, and ultimately design and conduct their own experiments.

With the development and acceptance of the overall curricular strategy, writing-to-learn assignments and writing-to-communicate assignments became the "how" of the three learning objectives that were associated with the writing assignments that taught critical thinking. To reiterate:

1. Students learn how to think more critically about the engineering content and learn discipline-specific ways of thinking (writing-to-learn and writing-to-communicate).

2. Students practice discipline-specific writing skills (writing-to-communicate).

3. Students learn that the process of writing and the process of thinking are linked (learned experientially from both).

The above model operated within approved theory for writing pedagogy, while at the same time worked within the existing departmental curriculum structure, taking into account that different courses present different contexts for teaching writing (Herrington [1985] 1994). Placing

Table 4.1. Examples of Assignments for Lower- and Upper-Division Courses

Writing-to-Learn Assignments (primarily lower-division courses)	Writing-to-Communicate Assignments (primarily upper-division courses)
• Written explanation of new engineering concept for a less-technical audience	• Peer-review memo of a technical report draft
• Collaboratively written brainstorming notes to promote group problem solving	• Letter or memo report of lab work written for a manager
• Prose explanation of how a quantitative homework problem was solved	• Formal technical report written for a scientific or research engineering audience

writing-to-learn assignments and instruction in the lower-division courses and writing-to-communicate assignments and instruction in the upper-division courses, in particular the senior project lab capstone course, mitigated resistance by taking into account available course and instructor time.

Class size and type also factored into the type of writing assignments developed for a course. Writing-to-learn assignments, in general, required less classroom time and less instructor time for evaluation and so were better suited for larger, lecture-based courses. Writing-to-communicate assignments, which are often longer and more complex, were better suited to a lab-based course that offered multiple classroom formats to facilitate the repeated cycles of feedback and evaluation normal to a process-based curriculum. The senior capstone course offered more opportunities to integrate lectures on technical and professional writing, as well as to restructure the syllabus to include time for writing process and support (including drafting, peer review, and consultations with the writing consultant). Table 4.1 lists some examples of writing-to-learn and writing-to-communicate assignments that are common to an engineering WID context.

HOW DID IT WORK? EXAMPLES OF ASSIGNMENTS AT BOTH LEVELS OF THE CURRICULUM

Writing-to-Learn Assignment: Explaining Critical Insulation Thickness
Contextually relevant writing-to-learn assignments made sense to the engineering faculty as we integrated them into lower-division engineering content courses. A particularly successful collaboration to develop a discipline-specific and low-stakes writing-to-learn assignment in a course about heat transfer resulted in a unique approach to students learning about a new concept via writing. After hearing a lecture about critical insulation thickness and completing homework problems to learn how to

calculate it, students were asked to write a brief, one-page explanation of the concept "critical insulation thickness" to an audience of high school students. Although the heat transfer students could assume their audience had a good understanding of geometry and trigonometry, the high school students would not understand the specific mathematics behind critical insulation thickness nor the terminology unique to heat transfer. In describing the new concept to a less knowledgeable group, the heat transfer students tested their own depth of understanding of the concept.

Once completed, the professor gave the one-page explanations to a colleague who taught mathematics at a local high school associated with the university. The high school students read the explanations of critical insulation thickness and marked the papers indicating where information was unclear or incomprehensible. They then wrote down questions that they would ask if they could speak to the author of the paper. The papers were returned to the university students so that they could see the degree to which their explanations had been successful or unsuccessful for the high school students. In general, the high school students commented little, which may have shown that they were able to understand less about the concept than the faculty expected. The comments, however, also showed that engineering students who used nonscientific (or nonexpert) metaphors were more likely to increase the level of understanding for the high school students. The university students, however, may also have needed to provide more accurate and detailed information to aid the high school students' understanding and to enable them to engage the concept.

Theoretically speaking, the assignment was not strictly an expressive, writing-to-learn assignment, because students wrote the explanations from within a subjectivity of expertise to a less expert audience. In addition, as part of the evaluation of the assignment, the course instructor required the identification of key concepts related to critical insulation thickness in student explanations. In this sense, the assignment was an extension of the problem sets students turned in regularly to the teaching assistant for grading and correction.

This assignment, however, is an example of an adaptation of writing-to-learn in a disciplinary context where conventional writing-to-learn assignments were culturally unacceptable. The assignment did maintain several of the conventional characteristics of writing-to-learn assignments. For example, the stakes for the students' writing were low, as the assignment was not worth very many points. In addition, the primary audience of the high school students would not evaluate the paper for quality or correctness, freeing the writers to take risks they might not otherwise take if they were writing to only a teacher-as-evaluator audience. The assignment also supported student learning in two additional ways: First, the high school students' feedback (or lack thereof) helped facilitate the heat transfer students' learning, indicating gaps in their own understanding or articulation of the concept. Students learned that they already took the most basic terminology of heat transfer for granted (terms such as "conduction," "convection," "radiation"), even though they would not consider themselves experts in the subject. Second, in preparation for the assignment, the heat transfer students had an opportunity to discuss the concept with peers in class in order to refine their understanding and to strategize about how to explain the concept. Students learned that by first talking the concept through with each other or the writing consultant, they were better able to write the conceptual explanation. Discussing and exploring a new concept orally is another occasion for learning that often supports writing-to-learn assignments.

As successful as the assignment was, especially from the perspective of faculty buy-in and collaboration, several areas for improvement became clear for the next implementation of the assignment: (1) students needed more direct instruction in strategies for explaining a technical concept, including the value of using nonscientific metaphors for less technical audiences; (2) in order to take the oral component more seriously, students needed more explicit explanation about the value of working a concept out orally before attempting to write it down; (3) more interaction with the high school students was needed so that they would have a greater investment in the project and provide more detailed comments; and (4) more time committed to the assignment would enable heat transfer students to revise their explanations based on the comments received in order to further refine their understanding of the concept and their ability to articulate it to a less expert audience.

Sample Writing-to-Communicate Assignment: Reporting Peer Review Comments

The curriculum structure of the department necessitated that writing-to-communicate assignments, and in particular professional documents, would be focused in the senior project lab capstone course. In the second (spring) semester of the capstone course, the professors and the writing consultant collaborated to develop a peer review memo assignment that would situate students in a proto-professional subjectivity that they could expect to inhabit in an engineering workplace. The idea was that students would learn how to formalize giving professional critical feedback to colleagues in the form of a common professional genre: the memo. The assignment asked students to trade with a peer their complete drafts of the formal technical report of lab work that they had been working on all semester, complete a peer review process (as guided by a handout with guidelines of what to look for), and then write up their comments and suggestions to their peer in the form of a memo. Students received points from the professor based on the completeness of their peer review, rather than solely on the content of the comments they made about their peers' paper. The writing consultant also read through the memos and made comments on the tone and quality of the students' comments so that in the future peer reviews they could be more supportive and directive of their fellow students' writing.

The peer review memo was designed to meet multiple objectives for teaching writing and critical thinking at the upper level of the curriculum: (1) students learned and practiced the form of a professional document, the memo; (2) students learned and practiced the process of peer review, which also has professional applications; (3) students were exposed to the work of their peers for the purpose of learning about other projects being done in the class and/or comparing peers' methodologies and standards with their own; and (4) students received feedback from their peers on drafts of their formal reports so that they could improve on a second draft.

The major challenge of this assignment was in how to educate and motivate students to do a thorough and substantive review of their peers' work. As is often the case in even less formal occasions when students review each other's work, students found it easier, and safer, to focus on marking issues of spelling, grammar and syntax of their peers' papers, rather than to address the issues related to critical thinking, such as comments on how to improve the discussion, analysis, argument, and organization of the paper. One approach to addressing this challenge was to provide students with direct instruction in the differences between higher-order concerns (HOCs)—such as organization and the quality of the analysis—and lower-order concerns (LOCs)—such as grammatical and mechanical errors (Sommers 1980). By giving them a vocabulary to categorize the different types of concerns that they were expected to address in their peer's paper, they were able to more fully engage the review process. This more robust peer review process challenged the quality of the critical thinking skills and the writing of both the author *and* the reviewer.

Overall, students, professors, and the writing consultant were extremely pleased with the usefulness and productivity of this assignment.

As stated above, the major challenges were encouraging students to be as honest as possible with their peers, while remaining constructive, and motivating students to engage their peers' papers at a critical level deep enough to be able to make substantive suggestions for improvement.

CONCLUSION

As has been explained and contextualized in this chapter, in chemical engineering we developed a curricular structure and pedagogical approach that negotiated the disciplinary norms and assumptions of the engineering faculty and the writing consultants with the purpose of collaboratively designing a theoretically robust, yet also relevant and practical, approach to integrating writing into the engineering curriculum. What we came up with was engaged students from their first year to their last year in a curriculum that held them accountable to engineering standards for learning key concepts and processes, while at the same time supporting their learning in how to write purposefully and successfully in an engineering context. In lower-division courses, low-stakes writing assignments helped students to explore their level of knowledge as well as identify important gaps in it. These writing-to-learn assignments encouraged students to think about audiences as readers and not strictly as evaluators. In upper-division courses, students increasingly encountered texts of more complexity and formality. The writing-to-communicate assignments required students to apply their knowledge of engineering, science, and math and place it within a larger community framework. The writing assignments focused on students not only being able to write a professional genre, but also to demonstrate the critical thinking skills required of engineers.

As our experience shows, implementing a WID program is as much about the politics of collaboration as it is about a theoretical or pedagogical endeavor. When the writing consultant has been asked to describe her role in the Department of Chemical Engineering, she has foregrounded any comments by saying that above all the job was one of diplomacy. For all the expertise in writing pedagogy and curriculum that she could bring to the project, and for all the willingness and desire of the engineering faculty to collaborate, all was null if we could not find common ground from which to work. This common ground, which we found in this case in the common value of critical thinking, was often not obvious and required the ability to successfully translate and openly communicate the disciplinary assumptions and language, as well as administrative structures and processes, of both the WID program and the Chemical Engineering Department. On all accounts, we were both successful and unsuccessful.

Much work has been done above to describe and justify an approach to integrating writing into an engineering curriculum that most writing instructors would now take for granted. One of the key lessons of any WID program, however, is that disciplinary assumptions do not necessarily carry across disciplinary lines, and foundational work has to be done in order to establish common ground. In retrospect, it is possible to see that a key source for resistance from the perspectives of both the writing consultants and the engineering professors was the lack of a common model for our collaboration. A model—such as that of Peshe Kuriloff's (1992, 96) five-stage model of Goal Setting, Inquiry and Self-Study, Creating a Context, Implementation, and Evaluation—would have created clear milestones, another common ground to further mitigate resistance. In practice, our collaboration did roughly adhere to this model, though in a more intuitive and less-linear fashion.

Joint goal setting, the first of Kuriloff's stages, was institutionalized near the end of the second year of the writing consultant's term when the Chemical Engineering Undergraduate Curriculum Committee adopted as a guiding document "Learning Objectives for Communication and Teamwork in ChE." This document outlined the developmental pedagogical model for an integrated four-year curriculum as described above. This was the first instance of the Chemical Engineering Department bureaucratically adopting and ratifying the language and theory of writing and rhetoric studies, a sign that some common ground had been found. In one sense, collaboration with the department had come full circle—from beginning with the existing general learning objectives in the department's *Strategic Plan* to the development of a new document clarifying specific learning objectives for written communication. Throughout the process of change and collaboration, the diplomacy of doing interdisciplinary work guided every stage. Without such careful attention to diplomacy, the challenge of change would have been insurmountable.

NOTE

Sarah was the writing consultant for the Department of Chemical Engineering. Through the space for diplomacy she created, the cooperation needed for curricular innovation was made possible.

5
MOVING TOWARD SUCCESSFUL INTERDISCIPLINARY INTEGRATION IN TEAM-TAUGHT COURSES
Building Cultural Bridges through Assignments

Mara K. Berkland

As colleges and universities examine and explore interdisciplinary teaching and learning, attention is also being paid to how instructors who team-teach can successfully integrate different knowledges and learning goals into courses, classes, and assignments (Graber and Pionke 2006; Hirsch et al. 2001; Penny 2009). As seen in the two previous chapters, one of the impediments to successful interdisciplinary teaching is disciplinary worldview socialization, which impedes comprehension, connection, and collaboration of not only students but instructors as well. In order to model successful interdisciplinary interactions for our students, as well as to effectively teach students how to integrate different ways of knowing into their learning, we must, as team teachers, work together to "create not only a unified outcome but also something new, a new language, a new understanding" (Fiore and Colarulli, 1997, S157).

Wrestling with these inherent challenges in interdisciplinary teaching requires deep reflection and ongoing conversations. As more than one interdisciplinary expert cautions, instructors of co- or team-taught interdisciplinary courses should be aware of the additional time required to execute such collaborations effectively (Darling 2011; Krometis et al. 2011). The result of the cost of interdisciplinary courses is that institutions and faculty members try to find creative solutions that engage multiple perspectives with little resource expense (McDaniel and Colarulli 1997). In extreme cases, these courses are little more than combined minicourses where two or more topics are taught separately under a single course banner, using a "divide and conquer" approach. Instead, interdisciplinary courses should blend assumptions and expectations by bridging knowledges in lectures, assignments, and exams. Instructors should mirror the integrated learning that they expect of their students by modeling the third cultural perspectives and practices that emerge from collaboration.

The purpose of this chapter is to demonstrate three teaching approaches—which I have adapted and term "application," "mirroring," and "analysis"—that instructors can use to artfully blend the multidisciplinary perspectives that emerge from disciplinary collaboration. These three approaches can serve as models for instructors to begin dealing with the difficult work of learning a different discipline, a different language, and even a different way of knowing to create courses that are truly multidisciplinary. To set the foundation for the discussion of these approaches, this chapter will first briefly review the literature on disciplinary worldviews and second explain how disciplinary worldviews can impede integrated teaching. Then, using the context of an upper-level engineering course as a case, the chapter will conclude with examples of application, mirroring, and analysis in practice.

WORLDVIEWS AND DISCIPLINARITY CREATE STRUGGLES FOR INTERDISCIPLINARY TEACHING

The "interlocking *system* of beliefs" that each discipline holds (Devitt 2004, 5) directly correlates to the teaching, evaluating, and testing methods that instructors employ in their classes. This form of understanding is referred to by a number of different names including worldviews, epistemologies, disciplines, cultures, and ideologies. Regardless of the specific term scholars use when referring to the expectations that frame disciplinary understanding, each is attempting to explain that when "any given discipline is dynamic and composed of different theoretical and methodological approaches, it will tend to share a language, a set of tools, and epistemological commitments" (T. Miller et al. 2008, 2–3). In other words, disciplines, much like cultures, privilege ways of thinking that guide the axioms and commitments that their members adhere to (Penny 2009). These differing worldviews have the possibility of creating conflicts when one interdisciplinary partner represents a social science or humanities perspective such as communication, rhetoric, or composition and the other partner comes from a discipline such as mechanical engineering, simply because the theories and the methods of these disciplines *appear* to have little to no similarity.

As Charles Snow (1959) posited as far back as a half a century ago, the divide between the sciences and humanities had emerged and was continuing to grow because of division and misunderstanding between two groups of scholars. As Snow (1959/1960) explains, "For constantly I felt I was moving among two groups—comparable in intelligence, identical in race, not grossly different in social origin, earning about the same incomes, who had almost ceased to communicate at all, who in intellectual, moral and psychological climate had so little in common that instead of going from Burlington House or South Kensington to Chelsea, one might have crossed an ocean" (3).

This segregation has evolved to the point that in some departments, colleges, and universities, it is assumed that humanities scholars and scientists are incompatible because of qualitative/quantitative and value/fact dichotomies. And, even if the divide had not at one time been so great, the reassurance that there was a division further fueled and reinforced such beliefs to the point that to date, scholars do, indeed, operate in separate spheres. The consequence of the divergent disciplinary evolution is the existence of unique, isolated disciplinary worldviews, epistemologies, ontologies, axiologies, and norms that make interdisciplinary interactions very challenging (Fiore and Colarulli 1997; Hirsch et al. 2001; T. Miller et al. 2008; Penny 2009).

One argument is that the sciences, in particular, have focused their research pursuits on attempting to understand the truth, hypothesizing about the truth, and proposing theories with the hopes of uncovering truth. In contrast, humanities scholars embrace the relativism of arguments, ideas, and perspectives (Devitt 2004; Palmer and Marra 2004). More specifically, the types of teaching and learning humanities and engineering scholars require of their students can vary greatly. Looking at the way communication scholars and engineering scholars approach teaching and lessons paints a clearer picture of what are considered valuable learning experiences.

Because engineering has at its roots a history of trades education, engineering programs have evolved to create a wide range of "technical and mathematical challenges" that span a variety of knowledge areas and that prepare future engineers to problem-solve while they interact with diverse populations (Booth 2004, 10). In their disciplinary discussions of pedagogy, they thoughtfully and thoroughly examine the ways that instructors can help students to understand the processes, systems, models, and patterns within their field of learning (Felder and Silverman 1988). First and foremost, students must learn concepts and then students must learn to navigate, apply, and even expand the complex models that these concepts come together to create. Engineering methods reinforce the application of truth and right answers with an emphasis on problem-solving (Palmer and Marra 2004).

By comparison, communication and writing scholars emphasize experience, independent processing, and creative theorizing (Penny 2009). Important to humanities and social science pedagogy is an "emphasis on historical and theoretical contextualization," where students are encouraged to explore the political and cultural context of knowledge (Penny 2009. 42). In other words, humanities scholars are encouraged to understand the values and viewpoints that help to interpret a situation or question a piece of evidence. Ideally, students are encouraged to understand and examine perspectives and evidence that help deconstruct an issue. In their assignments they are expected to raise debate, evaluate evidence, and contextualize viewpoints, which remind students of the uncertainty and variability of perspectives and right answers (Palmer and Marra 2004).

Despite the fact that first so much attention has been paid to the humanities/science divide, and, second, initially at least, these two areas seem to have different goals, scholars from a variety of disciplines argue that there is more space for intellectual and disciplinary overlap than our current, rigorous divides encourage (Douglas, Koro-Ljungberg, and Borrego 2010; Howe 2009; T. Miller et al. 2008). Thaddeus Miller et al. (2008) caution against maintaining firm disciplinary boundaries, explaining that when scholars (in this case researchers, but just as easily co-teachers) focus too much on their differences and continue to anticipate and work around, rather than bridge, the disciplinary divide, "there is minimal recognition of a larger, integrated system that exhibits complex relationships between the subsystems themselves" (3). In other words, even if we embrace and support interdisciplinarity and team teaching, we may still expect the above-mentioned differences and attempt to avoid conflict through a process of division and delegation. The consequence of an approach that divides and delegates is that we miss the opportunity to connect over the perspectives that we share, and create nuanced, integrated perspectives more similar to what our students will apply in their careers.

The perspectives shared by humanities and engineering scholars are illustrated well by looking at teaching goals. Anna-Karin Carstensen and Jonte Bernhard (2009) explain in an article on learning environments in engineering,

> Three concepts important to learning environments are the intended object of learning, the enacted object of learning, and the lived object of learning. The intended object of learning is the subject matter, along with the skills and values that the teacher or curriculum planner expects the students to learn. The enacted object of learning is the space of learning

> comprised within a learning environment, i.e. what is actually made possible for the student to learn. The lived object of learning is the way students see, understand, and make sense of the object of learning, along with the relevant capabilities the students develop as a result. An important object of learning in engineering education is that students should understand and learn to use theories and models and link them to objects and events. (394)

In other words, the goal in their classrooms is to improve students' ability to understand, apply, and negotiate concepts in the world around them. Although written for an engineering audience, Carstensen and Bernhard's (2009) concepts could easily describe the learning expectations of a writing classroom. The vocabulary might be different, and at first confusing, but after examination we can see that instructors, regardless of discipline, strive to create spaces where students can make sense of information and practically use that information in the world around them. Humanities instructors similarly want students to be able to critically think in order to responsibly interact in the world around them (Walker 2009). Regardless of the discipline, instructors want students to be able to apply knowledge and become independent thinkers, regardless of whether they might differently emphasize the place of creativity, inside a model or outside of a model; emotion and perspective; and the permanence of knowledge.

It is evident that similarities exist among disciplines. It is helpful to remember this during the moments when the disciplinary strategies or values that arise cause conflict in an interdisciplinary classroom. The work of learning and respecting the concepts or values that come from another discipline and that reflect the disciplinary concept or worldview differences is challenging, but these differences should not dissuade scholars from collaborating (Dillon 2008). A pedagogy that connects two distinctive disciplines has the capacity to generate "novel, original, and unexpected" learning moments (Dillon 2008, 255; Dillon 2006).

INTEGRATED PREPARATION LEADS TO INTEGRATED LEARNING

Because of the time and resources that interdisciplinary teaching demands, one of the more common models for team teaching places two instructors in a course or a course series and leaves each instructor responsible for checking, grading, or defining his or her contribution to the course (McDaniel and Colarulli 1997; Shapiro and Dempsey 2008). However, Elizabeth McDaniel and Guy Colarulli (1997) conclude that students are better positioned to make connections between or among disciplines when the instructors themselves have spent time considering the connections themselves in ways that shape the course design, course goals, assignments, exams, and lectures, which makes a strong argument for creating assignments and examination processes that reflect an expectation of integrated learning. The problem is that "designing assignments that invite integrative or interdisciplinary learning is an art form as well as a pedagogical skill that has yet to be discovered, let alone mastered" (*participant*, Lardner and Malnarich 2009, 33).

Connected or integrated teaching is more easily accomplished when interdisciplinary partners identify points of commonality and then work together to formulate the question, share methods, and identify problems (Alber 2001; T. Miller et al. 2008). If that step is omitted, it is difficult to create assignments and exams that effectively integrate disciplines, and students will then fail to make the connections that instructors hope they will. Once teaching collaborators have embraced this process of mutual learning and sharing, a step that is more thoroughly delineated in the first chapter of this book, the secondary step is the creation of materials that reflect a hybrid or integrated perspective. Emily Lardner and Gillies Malnarich (2009) provide a helpful explanation of what integrated teaching and successful interdisciplinary student work could look like. As they note, "integrative learning includes but is not limited to interdisciplinary learning" (Lardner and Malnarich 2009, 32). Interdisciplinary learning assimilates modes of inquiry specific to each discipline and is one kind of integrated learning. Lardner and Malnarich (2009) emphasize that integrated learning is more complex and encourages the inclusion of skill-based expertise, personal experiences, or ideas from disciplines outside of the ones being studied.

In other words, integrative learning is an "umbrella term for structures, strategies, and activities that bridge numerous divides such as high school and college, general education and major, introductory and advanced levels, experiences inside and outside the classroom, theory and practice, and disciplines and fields" (Klein 2005, 8). Further, integrative learning may ask students to draw not only on the disciplinary perspectives being taught in the course, but skills or experiences they may have developed in other courses that fulfill their curricular or general education requirements. As similarly echoed by Penny (2009), "There is a simplistic assumption abroad that to practice interdisciplinarity, one can simply drag the methodology or subject matter from an outlying discipline into one's own" (35). Instead, integrated learning is centered around learning multiple skills and theories and knowing when each one is necessary and how differing skills and theories can complement each other (Penny 2009).

Integrated assignments and courses emphasize the equal value of different disciplinary ideas, rather than starting with a complete course in one discipline and tacking on assignments to introduce another. The consequence of the add-on approach is less complex development of the secondary discipline, which is then treated as such, both in the course content and the minds of the students. However, equal value does not necessarily mean equal time or simultaneous execution. Successful interdisciplinary teaching produces integrated learning where the assignments or testing measures consider the disciplinary knowledge level and the learning goal of each lesson when integrating the disciplinary perspectives. As illustrated by Penny Hirsh et al. (2001) in their Engineering Design and Communication course, sometimes activities have a predominant focus of a disciplinary perspective. For example, when students are alone, working on design, their focus is solely on engineering theory and training. At other times, they are presenting or proposing their products or research results, and the emphasis is more strongly on communication theory. In other circumstances, there is clearer integration such as in the laboratory where students are developing group communication at the same time as they are focusing on engineering design and experimentation skills. The theory sets are separate, but integrated in appropriate circumstances (Hirsch et al. 2001). Also of note is the fact that students' levels of knowledge of the two disciplines may be different and may even be evolving at different rates. Junior-level engineering classes may be integrating communication and writing skills, but perhaps with less depth than junior-level writing and communication majors. The goal is to help students learn and apply theories from both disciplines at the appropriate level necessary for the course, to recognize how they work together, and to assess when those knowledge sets need to be applied.

THREE APPROACHES FOR INTEGRATIVE TEACHING

Integrating disparate disciplinary cultures into one course is possible with thoughtful, reflective approaches to course planning. The three strategies of application, mirroring, and analysis provide scaffolding for the integrated teaching goals set forth above. As a model for these approaches, I will draw predominantly from the assignments and examination styles that were developed in an upper-level engineering course to which I had been assigned as a WID consultant. The course was a year-long sequential course that focused on the study of mechatronics. The course—which introduced students to dynamic system modeling, instrumentation, sensors, actuators, and computer-based data collection in a laboratory setting—was a junior-level laboratory course. It was, at the time, the only engineering laboratory course that had an instructor from outside of the engineering faculty. This opportunity was presented to the faculty who oversaw the WAC/WID program at the university simply because the engineering instructor supported his students' writing development and felt that engineering writing needed to be developed at all levels.

One desired course outcome was that future mechanical engineers would be able to explain the goals, processes, and results of their laboratory research to different audiences. It was with this goal that we developed assignment and testing methods that embraced application, mirroring, and analysis. Each of these tactics employs a form of writing. Strong writing and communication skills are expected of engineers, who are expected to be able to relay important information about projects and ideas (Artemeva, Logie, and St-Martin 1999; Hirsch et al. 2001; Gider et al. 2012). Moreover, from a teaching perspective, writing can also "function as a conceptual tool for assisting students in analysis, interpretation, and communication of scientific ideas" (Chamely-Wiik et al. 2012, 503).

Application

When students are given the opportunity to apply knowledge to real-life situations or examples, they are exposed to the reality of complex, interdisciplinary thinking and are forced to learn to integrate a variety of perspectives into their conclusions or solutions (Herman et al. 2005; Hirsch et al. 2001; Liu, Lin, and Tsai 2010). For writing (and communication) instructors, it can be challenging to create application assignments that meet the many standards of integrated learning. As John Savery and Thomas Duffy (2001) explain, designing authentic tasks is a way to prepare students for the cognitive demands of the environment in which they will eventually be immersed and create opportunities for integrated learning. As they refer to science education they clarify, "we do not want the learner to study science—memorizing a text on science or executing scientific procedures as dictated—but rather engage in scientific discourse and problem solving" (4). It is within a discursive problem-solving process that multidisciplinary knowledges and skills can be integrated and taught and students can learn how to negotiate that process (Gider et al. 2012; Hirsch et al. 2001).

In the case of the mechatronics course, which was a laboratory course, application was a bit more challenging than in other courses where instructors might have required a project assignment in which students interacted with a client (either pseudo or real). Since both instructors agreed that the mechanical engineering students needed to not only discover the right answer or formula in each week's laboratory experiment, but also be able to explain how they did it, why it worked out the way it did, and why students had any struggles, we agreed that the traditional laboratory style of the class needed to be altered.

During the first week of class we explained to them that as engineers, their experiments and results would often be of interest to diverse audiences. We reinforced a theme that had been introduced to them in their introductory engineering courses, that they would have to be able to explain what they were doing to marketing managers, finance managers, machinists, lawyers, and salespeople; consequently, the students needed to be able to solve complex engineering issues as well as explain the relevant information about their processes to interested parties. We explained that their experimentation phases would often be the subject of scrutiny when they worked for organizations of any type, and consequently we were no longer going to be requiring laboratory reports from them. Instead, all of their laboratory experiments would be written up in the form of a memo. Students were given a sample memo (see Appendix 5.A) that prompted them about appropriate information to include in such a document. Students were required to explain what they did, how they did it, and any problems that arose in simple language that could be read and understood by a variety of people. In addition, every week during the course, the engineering professor and the WID consultant worked together to teach a different communication writing point, with the consultant providing the communication theory/skill and the engineering professor providing the example.

The second half of the course revolved around a robotics project competition where they met in teams and gave project updates each week. The project updates replaced the lab reports; again, they turned in memos that updated everyone—the team members, the teaching assistant, the communication consultant, and the course professor—about the important aspects of their progress. As an additional component in the second half of the course, students were expected to present their robot design to the entire class (and invited guests). The presentation made up a large portion of their grade, and students were prepared for the presentation through the corrections and guidance they received for their weekly project updates in their team meetings. Students were instructed to give a team presentation that highlighted the unique techniques that undergirded their final project (the robot) and any unique design features. The presentation was required to be short, only ten minutes, because we wanted students to have to think about the main points of their presentation very carefully. We wanted to prevent students from talking about each step of their project, as the audience would not be interested in that much detail, but instead we wanted them to think about what to highlight and explain so they stood out, regardless of how well their robot actually performed in the final competition.

This application-based approach helped students understand how knowledges work in action. In a real-world circumstance, people are expected to present complex, reasoned, and multifaceted explanations of projects and issues. Performance and products are rarely segregated based solely on expertise. "After all, the students will be faced with technical matters on a daily basis in their careers. But quality writing is often more than the effective compilation of technical information, especially when the audience may be a group of investors, or made up of marketing personnel, politicians, even the CEO. Further, as our world is changing, there are ambiguous social and ethical issues to consider" (Piirto 1996, 307–8). Application assignments allow students to apply all of their skills and knowledges simultaneously, just as they will in their future careers.

Mirroring

The memo-writing assignment that stood in for a weekly lab report would not have worked well if the professor, teaching assistants, and the WID consultant had not also adopted the memo format as their primary form of communication with the students (first term) and teams (second term). As the research clearly shows, students will interact in ways that mirror the communication and interaction patterns of their course instructors (Webb, Nemer, and Ing 2006). This cognitive apprenticeship, as it has been termed, is a method in which teachers bring problem-solving processes and methods to students' attention so that students can study and mimic them (Collins et al. 1987). When instructors realize the opportunity to serve as models or master to the student-apprentice, the student then has a greater opportunity to "observe, enact, and practice [appropriate behaviors and thought processes] with the help of the teacher and other students" (Collins et al. 1987, 6).

The mirroring or apprenticeship approach was what we used with our assignments in the mechatronics course. After each of the weekly memos had been evaluated, the WID consultant, teaching assistants, and engineering professor would get together to discuss the problems with the assignment. Together, the group would draft response/correction memos, explaining to them what the major problems in understanding were for the group as a whole. Memos were used to discuss corrections of the communication issues *and* experimentation and design issues that arose in the experiments. A sample of a memo response crafted for a particular set of writing issues might cover format, language, organization, or content quantity (see Appendix 5.B).

Analysis

The final recommendation that many interdisciplinary teaching teams can implement into their courses is utilizing a form of examination that relies on narrative questions. Narrative questions can take similar formats as traditional multiple choice or true/false questions, but they demand a more complex, nuanced integrated understanding of the course material. Mertler (2003) gives a very basic example of a complex multiple-choice question from a single discipline:

> 1) A man stands at the center of a circle. He measures the distance to the edge of the circle as 3 feet. What is the diameter of the circle?
>
>   A)   3 feet
>   B)   6 feet
>   C)   9 feet
>
> 2) Explain your response to question 1 above. (176)

Fayyaz Ahmad Faize, Mohammad Arshad Dahar, and Asaf Niwaz (2010) provide a more complex multiple-choice question from their research on student learning in physics. In this question, students are given a question, are offered four options in a multiple-choice format, and are expected to give their justification for choosing or rejecting each option.

Question: What **must** change when a body is accelerating?

Option	Reason
A. the force of acting on the body	Wrong answer: force acting on a body may be constant and still the body will be accelerating e.g. a falling body with no air resistance
B. the velocity of the body	Right answer: acceleration is the change in the velocity with time, thus velocity must change to produce acceleration
C. the speed of the body	Wrong answer: A body moving in a circle with constant speed do [sic] have acceleration called centripetal acceleration
D. the mass of the body	Wrong answer: the mass of the body may increase or decrease and still it may not produce acceleration e.g. a body at rest and changing its mass. (Faize, Dahar, and Niwaz 2010, 205)

Unlike question formats that rely solely on memorization of information, such narrative exam questions as the ones above demand metacognition. In these cases, students are forced to make thought processes external, which then makes them "available to both the student and teacher for observation, comment, refinement, and correction" (Collins et al. 1987, 4). This process of making external what is normally internal—in this case, the logic and thought processes behind decision making—is useful for testing the integrated information of interdisciplinary courses where students are required to simultaneously apply multiple perspectives, knowledges, and disciplines to reason the answer. Such complex styles of examination are more in line with our education objectives as teachers and more adeptly guide students through the multiple levels of learning that we require of them (Collins et al. 1987).

In addition to complex thinking, many courses, but particularly in interdisciplinary ones, instructors struggle to help students understand the range of perspectives an issue might engage. There is a certainly a time element to this problem, but narrative exams are a simple way of moving students to a more complex level of thinking. According to Thomas Terry (1980), narrative exams benefit instructors because such question formats allow for a more complex, developmental integration of the material in ways that traditional examination styles (multiple choice, true/false, essay) do not. As Terry (1980) notes, "Often we cover a variety of topics and principles without providing students with a sense of how all these topics interrelate in nature" (157). The narrative exam option creates an opportunity for the instructor(s) to encourage thoughtful analysis and evaluation where students have to reflect on and explain their conclusions about a complex issue.

What might an interdisciplinary narrative exam question actually look like?[1] In the case of a mechanical engineering course, a question might look like (the question below provides only two answers that serve as illustration, but a typical exam question might have four or five options to choose from among):

> Question: You are preparing a presentation to clients of a company that has hired your team to develop a production line for their line of airplane valves that has a very small margin of error in the manufacturing process. You have been working on the quality control leg of the project, specifically the machine vision (MV) component of this project, and you are going to be late on a deadline. What might you tell them about your progress? Of the five options below, only one is the best answer. Note first whether the answer is the one you would choose, and justify your response.

a) there has been a problem with the CCD, which is incredibly important because in a CCD image sensor, p–n junctions (essentially photodiodes) are used to absorb photons and produce charges representing sensed pixels, and the CCD is used to read out these charges. Once we have the CCD issue resolved, we hope to move quickly.

Hypothetical Student Response: I would not choose this one. There are two problems with this answer. First, the audience is a group a clients, and it can be assumed that many of them are not engineers. Consequently, this statement employs way too much jargon. Words such as "CCD image sensor," "p-n junctions," and "photodiodes" need explanation and should only be included if such knowledge is necessary or the audience is full of engineers. Second, CCD sensors employ reverse-biased p-n junctions.

b) Acquiring an image in machine vision is just like taking a picture—getting an accurate representation requires just the right amount of lighting as well as the right lens to capture a picture of the object and pass the information onto the sensor. A typical machine vision camera has a sensor with approximately 500 × 500 pixels, but because we want to reduce the margin of error, we want a crisper image that will require much greater resolution. This is currently presenting problems for us.

Hypothetical Student Response: I would choose this one. This response is jargon free and is accurate in the information given to the client. The provision of the analogy provides an explanation that is clear but simplified. The average is 500 × 500 pixels, but in some cases where accuracy is incredibly important, the pixels can increase to something like 4904 × 3280 pixels. Of course, this has its advantages as well as disadvantages, but for a project like this more resolution is necessary to ensure accuracy.

In both cases, there was a mechatronics and writing component that needed to be addressed in order to fully answer the question. Narrative multiple-choice questions such as the ones above provide students with the opportunity to integrate different disciplinary knowledges just as they would in their professional lives, simultaneously. Such opportunities push them to analyze information in context and appropriately evaluate that information from more than one disciplinary perspective.

SUMMARY

Foundational courses in the humanities and the sciences approach knowledge differently and so often teach students to most obviously learn different information, but also regard information and its purpose differently. In cases where students are socialized into only one discipline, learning processes present little conflict as eventually all disciplines bring their learners into complex conversations that debate ontology and epistemology. For the undergraduate learner in an interdisciplinary course, or in a course or program that is taught by instructors from different disciplines, assignment descriptions or examination questions that integrate multiple ontologies and epistemologies may serve to frustrate and disengage students who might otherwise contribute and excel. In such circumstances, it is the responsibility of the instructors to develop evaluation materials that integrate the seemingly competing frames of fact and value and show students early in their learning how these forms of knowing work together. Ideally, exams and assignments ask students to show mastery of the evaluation processes and truths of each discipline through application questions that engage students in real-world problems.

By employing three simple techniques, collaborating instructors can easily cross boundaries to integrate different knowledges into assignments and exams. Such assignments avoid the "divide and conquer" approach to interdisciplinary teaching and instead push student knowledge up the ladder of learning, where they can simultaneously apply and evaluate the different knowledges demanded within integrated assignments. By using co-constructed assignments that employ application, mirroring, and analysis, interdisciplinary team teachers can create a unique course that mirrors the issues, problems, processes, and discussions that students will actually find in the real world.

APPENDIX 5.A

Memo to Students Regarding Laboratory Memo Assignments

MECHANICAL ENGINEERING MEMORANDUM
TO: Mechatronics Students
FROM: Communication Consultants
SUBJECT: Laboratory Memo 2
DATE: X/X/XXXX
CC: Professor X, Laboratory Instructors

The purpose of this memo is to provide an overview of the guidelines for writing the laboratory updates that you will be submitting in place of traditional laboratory reports. Included in this memo is an overview of expected memo content as well as some general writing suggestions.

MEMO CONTENT

The memo should orient the reader to the specific issue or issues that are relevant to recipient/audience for the laboratory you executed. The memo should have an introduction, a body, and a summary.

Introduction

The introduction is brief and serves to explain the scope of the memo; in this case the memos will focus on the lab experiment you executed and what relevant observations emerged from the experiment. All memo introductions have a clear and brief preview statement explaining how the review is organized and what it will include, which will be the method, results, and discussion of the laboratory exercise.

Two of the most common mistakes in the introduction concern, first, telling the purpose of the laboratory exercise instead of the purpose of the memo, and second, neglecting to state the areas that would be addressed by the memo. Do not be tempted to start with a statement analyzing what the instructors intended students to take from the lab, such as:

> *The purpose of the Lab View Computer Systems and Computer Data Collection lab was to gain familiarity with the visual programming language Lab View and how it can be used to gather and analyze data in various applications.*

Although this is informative, it does not address the reason for the memo. An effective introduction should address the purpose of the memo and the specific issues it will address like this:

> *This purpose of this memo is to present the basics of data acquisition, measurement, and analysis utilizing LabVIEW. It will describe the methods followed in labs 2 and 3, as well as the results of those labs and any observations or applications.*

Body

Laboratory memos will have methods, results, and discussion sections. These three sections have distinct goals and limitations. Each section should be labeled clearly.

Method—WHAT YOU DID

This section explains what actions were performed during the experiment. It helps to imagine that you are writing this section as a checklist before you do the experiment, as if it were a recipe. If, for example, the method requires that you make adjustments during the experiment, tell us what those might be and how you will know to adjust.

Results—WHAT YOU FOUND
This section is brief. It only contains what you found. It should not contain any explanations for or interpretations of the results. This section should be limited to the most important things you observed.

Discussion—WHAT YOU THINK
This section, in contrast to the results section, is relatively long. It should contain all of your thoughts and conclusions about what you observed. It should extend the theoretical connections to relevant issues, particularly those issues that you are studying in class.

Conclusion

The conclusion is your last chance to remind the reader of what was said in the memo and to highlight any important aspects of the experiment. (Tell them what you told them and what was important about it!)

Writing Guidelines

When writing a review, there are some very important guidelines regarding point of view, information specificity, language clarity, and visuals.

Point of View

The debate about the appropriateness of using passive voice in scientific writing is still ongoing. What most technical writers agree on is that laboratory procedures should be written as first-person or third-person phrasings. For example, "We measured the distance" or "The distance was measured" or "The laboratory partners measured the distance."

Information Specificity

Writers need to explain and expand upon issues that they introduce. (If you do not want to expand on a point and the information isn't absolutely necessary, do not bring it up.)

> *E.g., The programs used in these labs will be similar to the ones needed for the downhill skier to operate. HOW? WHY? Are they similar? HOW? WHY? Are they different?*

Language Clarity

Technical writing should be as precise in its word choice as possible. If an idea can be said with one word instead of five, choose the one.

> *E.g., While exploring the photosensors, it was found that they are of very good use but not extremely precise devices.—Wordy*

> *E.g., The photosensors were useful but not precise in their measurements.—More concrete*

SUMMARY

Keeping in mind these stylistic and content guidelines for laboratory memos, mechatronics students should be able to successfully explain, discuss, and justify their projects.

APPENDIX 5.B

Memo to Students Regarding Laboratory Memo Corrections

MECHANICAL ENGINEERING MEMORANDUM

TO: Mechatronics Students
FROM: Communication Consultants
SUBJECT: Laboratory Memo 2
DATE: X/X/XXXX
CC: Professor X, Laboratory Instructors

The purpose of this memo is to highlight some of the difficulties in the second mechatronics laboratory memos. This memo will examine how to use equations in the texts. The examples and guidelines that I will cover conform to *APA Publication Manual* standards.

EQUATIONS

To display equations, the following guidelines should be applied:

- must have a number in parentheses at the right of the page;
- must be numbered in the order they appear;
- must be able to be read as part of the text;
- must be referred to using a capitalized name (Eqn. 1);
- must not include *, **, ^, E, or other "computer language" usage;
- must have all symbols defined, either in previous text, in text immediately following the equation, or in a table of symbols.

Equations are like figures in that they assist the reader's understanding of your methods or analysis, so they must be easy to read and visually accessible. Generally, you would define the symbols used in the equations in either a table in an attachment or appendix, or after the equation. For the ME 3210 labs, you should consult your instructor as to whether s/he wants you to do this.

Example

Then using this data and the following equation, Re was found (see Eqn. 1).

$$Re = r \, vL \, / \, m \qquad (1)$$

Where r is the fluid density, v is the fluid velocity, L is a reference length (usually diameter for pipes), and m is the fluid viscosity.

SUMMARY

When using supplementary information such as equations, it is important to make them visually accessible to the reader. Equations and figures must be clarified and explained in the text—they can't stand alone.

NOTE

1. I have developed my analytical or narrative problem approach more recently by working with my co-teacher in an interdisciplinary course entitled Human Sexuality. This course, co-taught with a biology professor, seeks to integrate the perspective of gender studies with the biological study of human sexuality.

6
"I DON'T HAVE TO ARGUE MY DESIGN—THE VISUAL SPEAKS FOR ITSELF"
A Case Study of Mediated Activity in an Introductory Mechanical Engineering Course

Maureen A. Mathison

In recent years it has become more common to incorporate extended writing assignments into curricula across the university. To that end, we might find instructors in biology lecturing on how to write laboratory reports, just as we might find students in sociology conducting peer reviews on their final drafts of a book review. Many students, however, find it confusing to enroll in composition courses and at the same time receive instruction in their own discipline about what counts as "good" writing. In their first-year composition courses, they are learning general rhetorical principles that they can apply to many writing situations. In their more specialized courses, they are learning the rhetorical conventions valued in their discipline. At times students find the general rhetorical principles and the rhetorical conventions of their disciplines in tension. They have difficulty understanding how the two can be complementary rather than at odds.

In my recent work in the discipline of mechanical engineering, I have found that often students do not recognize the importance of written prose, primarily because of the emphasis on the teaching of the numerical and the visual in their more technically oriented classes. Writing is generally seen by such students as something that is done "over in the humanities." While arguing is said to play a critical role in the engineering curriculum (and is often mentioned in the syllabus), students receive little or no instruction regarding its relevance and application to the work they will do as professionals. Thus while instructors often think of themselves as teaching argument, they are actually teaching content without the rhetorical strategies through which that content might be given significance. This chapter examines more fully how argument is

DOI: 10.7330/9781607328032.c006

situated in a technically oriented discipline. Specifically, I examine the teaching and learning of argument in a first-year mechanical engineering course at a public university in the western United States as students are introduced to some of the basic principles of communicating in their discipline, and as they interpret what it means to argue as an engineer.

ARGUMENT IN MECHANICAL ENGINEERING

Argument as it is taught in engineering classrooms is distinct from the traditional views of argument often taught in composition courses. In composition courses, what comes to mind when we think of argument is the supporting of claims with reasoned evidence. The discipline has generally adhered to Stephen Toulmin's model of "data, claim, and warrant" (1958, 97–99), teaching students that a good argument is one that asserts a claim based on logical reasons that are presented through language.

In engineering, or at least in mechanical engineering, what counts as "data, claim, and warrant" differs, as language is often not considered the primary mode of knowing. An engineer's way of knowing involves multiple symbol systems, some of which are verbal and some of which are numerical and visual. More often than not, they rely on the latter to make a claim persuasive; calculations and schematics often dominate their texts. In addition, the terms under which an argument is judged logical is generally not inferred but is often considered a priori through the knowledge that is applied at particular points in the text. Toulmin (1958) explains, for example, that "In his *Philosophical Essay on Probabilities*, La Place draws explicit attention to this class of substantial-yet-conclusive arguments: 'In the applications of mathematical analysis to physics,' he says,' the results have all the certainty of facts'" (137). In such cases, the writer must contend with the very validity of the theory applied to a problem dealing with, for example, thermodynamics, or with the application of a calculation, such as Ohm's Law, in order to question or dispute the conclusions on which a claim related to them is based.

Many of the more technical disciplines operate out of this argument field, valuing the theoretical and its nonverbal forms of meaning making as a method to persuade. It is important to examine argument in this context because many of the students who enroll in higher education composition courses are in fields of study that require they learn to argue through multiple means, emphasizing the numerical and visual over the verbal as the primary means of persuasion, without recognizing the importance of the verbal.

The perceived uses of these means of argument lead to different ends, one being more text based and the other more product based. In composition courses students learn to understand the rhetorical conventions associated with texts in their disciplines and to present information coherently and logically. The final product is emphasized in written form, whether it be a summary or a research report. In the more technical disciplines, such as mechanical engineering, the pedagogical goal is to teach students to employ the principles of science to technical problems and to solve them. The final outcome is often emphasized as, quite literally, a product or a prototype, a three-dimensional model of an object—a grease trap, a ski apparatus designed for paraplegics, or even a device to improve upon cancer radiation treatment.

Working with students in the more technical disciplines can be frustrating for someone such as myself whose own discipline tends to value the written over the numerical or visual. A curriculum that supports technical thinking is based on physics and mathematics, courses in which the written is generally considered after the fact. It is common for students to believe that writing is not integral to the design process (in which students conceptualize and build preliminary models of a mechanical device), but is ancillary. Students consider it redundant because after all, the visual itself embodies the underlying principles and illustrates their technical application. Resistance to the written can be heard in such phrases as "I don't have to argue my design—the visual speaks for itself."

Such comments point to what Bucciarelli (1996, 62) has termed the "object world," the conceptual modus operandi through which engineers design and construct products. According to Bucciarelli: "it is the fixation on the physics of a device that promotes the object as icon in the design process . . . while different participants in design have different interests, different responsibilities, and different technical specialties, it is the object as they see and work with it that patterns their thought and practice" (4).

Many students learning to become engineers remain fixed on the device itself, ignoring the rich social processes by which its design is influenced. They focus on the inconsistencies and weaknesses of it at the expense of the actual communicative processes that help identify and correct them. Students in the technical disciplines often have a tendency to believe that if they just apply the right scientific principles and technical know-how, they can either create or improve upon a device. Ultimately what is at stake for them is a product that works, not a final research report that describes it.

Unfortunately, this belief oversimplifies the everyday communicative conditions in which many engineers work. Mechanical engineering is a discipline that requires interaction among the many people who have a real interest in the outcome, from the design team itself to the prospective client who might purchase its product. While it may seem that the appropriateness of communication in these varied situations differs—with some taking place primarily through the numerical, others primarily through the visual, and still others primarily through the verbal—they generally operate simultaneously. For example, individual members of a design team contribute their unique expertise to a project, and therefore, while one engineer may be working on one aspect of a device, a second may be working on another. At some point the two must come together to examine how the features of the one aspect influence the behavior and performance of the other. In this way, necessary changes to the design can be determined. The engineers communicate through a variety of means. As the engineers talk, both take notes. Those notes may take the form of mathematical formulas, visual schematics, and verbal jottings. After the meeting, they may continue their discussion through email and pass on their decisions to others on the team through memos. Note that multiple symbol systems are in play to generate and modify ideas as the concept is revised and argued (see Winsor [1992] for a discussion of what counts as writing in engineering).

Many students, however, do not recognize the scope of communicative practices involved throughout the design process, with written prose being the least acknowledged by them. The results of one study showed that while "research and writing occupied an average of 35 percent of team members' time; oral interaction occupied more than 46 percent; and the activities typically associated with engineering design—sketching, model building, and analysis—occupied only 17% of their time," students attended very little in their group discussions to the production of written texts (Geisler 1993,174). This lack of attention conflicts with actual workplace practice, where the responses of seventy engineers reported that they allocated an average of 33 percent of their daily time toward producing written prose. Claimed one retired engineer, "I spent 25% of my day writing, which was worth 50% of my value to the company" (Mathison and McCulley in preparation). How is it then that students do not understand the importance of written prose in their discipline?

CLASSROOM ACTIVITY AS A LENS INTO THE TEACHING AND LEARNING OF ARGUMENT

One way to understand how students develop certain beliefs about writing and its relevance in their discipline is to examine the contexts in which they learn about it—the classroom. Research examining classroom discourse has pointed to the powerful ways in which instruction helps shape students' thinking about disciplinary information (Lemke 1993; Mitchell 1992). This approach locates the teaching and learning of disciplinary ways of thinking and behaving in the culture of the classroom, with the instructor as the modeler and mediator of professional beliefs and practices. The tasks or activities instructors assign students are those they believe will assist them in learning about the discipline.

The *activity* of the classroom, then, as it is embodied in the goals of the syllabus and enacted through discussion and assignments, are examples of what James Wertsch (1994) might call "carriers" of sociocultural patterns and knowledge (203), providing students a type of social and intellectual guide to performing successfully in a discipline. Through these activities, students begin to understand what information is important and how to apply it to relevant situations. At the same time, these activities are mediated through students' prior experiences, understanding of, and disposition toward the material, and particular needs given the context of their application and use.

In the remainder of this chapter, I report on the findings of a case study of one undergraduate classroom—an introductory design course in mechanical engineering—to examine how it is that students might come to believe that "the visual speaks for itself." In particular, I am interested in the manner in which different symbol systems (the numerical, visual, and verbal) are situated in the day-to-day activities of the classroom and how students perceive written prose in the design process. In short, this research investigates how beginning mechanical engineering students are taught *what* communicates and *how* it communicates, providing a lens into how argument is presented to them.

The class observed in this study was the first of six design courses being revised to incorporate writing and speaking in a mechanical engineering program. The fifty students enrolled in the course were not yet accepted into the degree program and were considered "pre-engineering" students. The goal of the course was to provide students a full, but bounded, experience of what it entails to be an engineer. According to the syllabus, the purpose of the course was to introduce students to the "overall process of engineering design through the use of lectures on methodology of the design process and through the experience of designing, constructing, testing, and competing with a mechanical device . . . This experience is designed to give [students] a perspective on the design process that will be giving [them] the mathematical and physical knowledge and skills to become a better engineering designer and problem solver. In addition, [they] will learn and practice the techniques necessary to communicate [their] ideas to others in an efficient verbal and visual manner."

To achieve this type of experience for students, the course was divided into three units: (1) lecture (homework, quizzes, and final examination); (2) laboratory (laboratory reports) and; (3) the team design project (goals and scheduled deadlines for completing the design and testing of an actual product). Each unit provided unique activities for the purpose of different outcomes, implicitly related, but not explicitly taught. For example, the lecture was used primarily to teach students about technical drawing, the laboratory was used primarily to teach students to use an engineering software program, and the project was used primarily to teach students to build a prototype of a device in response to a problem. The first two units had a different instructor. Students worked independently in teams, without supervision, to complete the project. All of the major writing assignments in which students argued for the soundness of their design were assigned in this unit.

To determine how the course was a "carrier" of engineering patterns and knowledge and how argument was situated within classroom activities, I examined the syllabus and handouts. To further my understanding of the course, I attended all the lectures and laboratory sessions, and took notes. First, I took notes as if I were a student enrolled in the course learning content. This encouraged me to understand how mechanical engineers might think. Second, I took notes describing how writing was used and referred to in the classroom and lab. I analyzed the field notes to determine how communication was situated in the context of teaching. (For the purpose of this study, particular attention was given to language that referred to written communication about any aspect of the design process.) In addition to the course materials and field notes, a survey was given to the students to determine what they thought the class was about and where and how written prose fit into the design process. This second part of the study, then, examined what students may have learned about communicating their design, that is, how they argued their views about the design with their fellow engineers.

WHAT COMMUNICATES IN ENGINEERING DESIGN AND HOW IT COMMUNICATES

Although the instructor stipulated in the syllabus that the course was about the design *process*, it really was about the design *product*. Unlike Bucciarelli's object world where the device is being *designed*, this instructor's object world focused on the device being *built*. On the surface the purpose of the lectures was to introduce students to the foundations of engineering visual design, teaching them about the requisite technical skills in the discipline and providing them a common language with which they could talk about them. It became apparent, though, in looking at the distribution of information across the three units and in analyzing the discourse of the classroom over the term that the design was already assumed in the lectures. By this I mean that students were not explicitly taught how to construct a device, but rather were taught how to communicate an existing one through visual and numerical forms; verbal forms were generally omitted. In this classroom, communication meant communicating the parameters of a mechanical device to someone—a machinist—who could then build it. It was not about persuading someone to adopt a novel approach to the design; nor did it serve to have students argue that their design was technically the best.

This pattern of meaning began early in the term as the instructor introduced students to some of the very basic design skills they would need as engineers. Included in the list was communication: (1) technical, (2) communication, (3) analytical, (4) modeling, and (5) interpersonal. The instructor further broke communication down into seven categories, explaining to students that all were "jobs you have to do":

1. sketching;
2. technical drawings;
3. computer models;
4. oral;
5. written;
6. shorthand;
7. mathematics.

Although listed fifth on the list, written communication received scant attention, as it was nested in the discussion of the laboratory sessions. Ironically, the majority of information regarding the laboratory sessions referred to Pro-Engineer, a computer program, which given the proper input, could produce a schematic or visual of a design within minutes. The little reference given to writing assignments was in regard to their use as deadlines for the team design project. Writing, then, was not being presented as a means to argue, but of keeping on task. All the instructor said about writing was that written reports "help boil important ideas down. . . . [and should follow] common general rules, like spelling." Missing was any detailed mention of the design proposal or final written report, which constituted two of the major laboratory assignments. Smaller laboratory reports were required on a weekly basis as students wrote on the progress of their design. These reports were not mentioned either.

To summarize the class discussion of communication, the instructor drew a diagram on the board of the design process that moved hierarchically through what is generally considered a design process from the [customer] needs, to brainstorming, to refining ideas, to modeling, analysis, and simulation, which then culminated in a final design. Many of these stages were directly related to the skills that the students were expected to learn, though they were not discussed as a process but rather as a path, moving from one stage to the next. Written communication was not incorporated into this diagram and, therefore, was not modeled as part of it. If a hierarchy was beginning to emerge in terms of the value of skills, the technical would come first and the written last.

As the term continued the pattern of classroom activity supported this hierarchy, and communication became more narrowly defined as technical sketching. Written prose became less valued through its absence in classroom lectures and through its implied status when it was included. For example, the instructor told students one day that "most people in this room are left-brained and need to tap into the right brain." He continued by saying that an articulate communicator in mechanical engineering would exhibit the following characteristics:

nonverbal
synthetic
concrete
analogic
nontemporal
nonrational
spatial
intuitive
holistic

Seemingly innocent, the list includes many words that are counter to the nature of writing—nonverbal (versus verbal); nontemporal (versus narrative), nonrational (versus logical), and intuitive (versus purposeful)—and negates the value of argument through the verbal. Students were being taught to reorient their perspectives in order to view the world from an engineer's lens, which drew more from the right brain than the left, the hemisphere of language.

After the third week of the course, the daily lectures were actually not lectures at all but exercises from the textbook that were completed during the class period, with the instructor explaining the particular concept under study. As the class sketched together, the instructor would elaborate the necessary skills to produce it, such as using a grid accurately or constructing appropriate line types for their unique significations. More important, students learned to read a pictorial in one form and construct the same object in another. They were becoming adept at translating, moving back and forth among the different views that would provide information about an object. The goal was to articulate a conceptual world through the visual.

The visual began taking on a life of its own, speaking for itself, as witnessed in the following phrases spoken by the instructor throughout the term:

- The dots aren't part of the communication.
- You need to get your point across and not be artists.
- I do this so people can see what I'm talking about.
- If you can, get the information across in one view.
- People shouldn't have to assume what we mean.
- If you're going to learn how to communicate efficiently, to convey how something might look like, you have to learn some of the rules.
- You're gonna take a plane and tell somebody about it.
- On the back side [of a pictorial], the same sort of behavior is going on.

Students learned that they did not need to argue that the design was worth constructing, but rather that the design *must* speak for itself, must communicate its purpose and form. Communication then can best be summed up as it was presented in the lecture by the following statement: "If there's some way to misinterpret something you've put on paper, they [machinists] will." Ultimately, the instructor was not teaching students about the design process but was providing them a common visual language that would bring a design to fruition as a tangible product. Drawing was about communicating the behavior of an existing device to another, in this case, the machinist. Said the instructor, "In the real world you can pick any view you want, project off of that, and present it. If it's not presented accurately, the machinist is gonna want to know what you really want. Think about how it's [product] gonna be made, and give the machinist enough information to show them what you want." In other words, it was assumed that students had a viable design and did not need to persuade anyone of their decisions along the way; nor did they need to persuade anyone of the quality of their product. Build a three-dimensional prototype, not an argument.

STUDENTS' UNDERSTANDING OF THE WRITTEN

The survey responses showed that students overwhelmingly believed that the class was about the visual. They claimed that the course was, in the words of one student, "mainly a training class. I have been taught the basics of drawing and communicating strictly through pictures and models. I have also been given a taste of what it is like to be an engineer—i.e., starting with an idea and drawing pictures, presenting the idea and finally constructing a product."

While all of the students who completed the survey emphasized the visual aspects of course content, only four mentioned that writing was part of the design process. That is not to say that students did not think about the role of written prose. It was, after all, incorporated into the curriculum at various points throughout the term, though it was never explicitly taught. Survey responses to the role of written prose in the design process were varied, as seen by the language students used to describe it. Below are categories of responses with the number of student responses in parentheses (thirty-eight students responded to the survey, but the numbers exceed that because some of them explained that writing played multiple roles). In descending order, writing is used in the design process to

- describe (8);
- gain approval (7);
- record information for the purpose of a trail (6);
- revise a design (6);
- inform (5);
- understand (4);
- present a finished product (4);
- communicate at a distance (3);
- visualize thinking (1);
- organize ideas (1);
- articulate information better (1).

If we examine the language students used to elaborate their responses, we see that most of them view written prose as ancillary, as something done once the design is determined and a potential product exists. Half of the students (nineteen) remarked that writing occurred after the design was decided. These students were apt to claim that it was for purposes of describing, gaining approval, or presenting a finished product. Fourteen of the students claimed that writing occurred throughout the design process. One student claimed that "writing is important at each stage to help present, remember goals and steps taken, to allow others to sit down and in their own time review your ideas." In this case, the design process becomes more manageable, and the design itself is thought of as something to be revised and improved upon. But the majority of the fourteen students still found written prose to be peripheral to the process, remarking, for example, that "written reports must also be given to people in charge on a regular basis [to inform them of progress]." In cases such as this, the design was assumed and writing served to apprise parties involved in its development.

Four students said written prose played a role before the design was decided and after it was decided. These students thought it important to communicate with a client to determine customer needs and then again once the design was determined—to communicate back to the client, as well as to the company. Finally, one student claimed written prose was not important because engineers spent the majority of their time talking.

INCORPORATING ARGUMENT INTO THE CLASSROOM

As an introduction to engineering practice, students in this classroom were presented with the charge of visually communicating and, more specifically, with communicating a product rather than a process. In this classroom, the product was already presupposed; it was not contested as it would have been in the professional world, where engineers routinely submit written arguments that attest to the quality of their product. These students did not have to engage in argument. In this case, they assumed their designs would result in a quality product, one whose features spoke for themselves. As a result, the importance of written prose in the design process became subjugated to the visual. This attitude may have been subtly reinforced throughout the term as the instructor emphasized right-brain activities at the expense of left-brain ones. The idea of argument became associated with the visual rather than the verbal. Earlier I mentioned that engineers use various means to engage each other in thinking about and arguing for particular aspects or features of a design, yet in the classroom they were being trained to focus on only one. As a result, their developing sense of what it means to be an engineer, to interact in and produce written documents for an engineering environment, was limited. An analysis of their writing showed, for example, that there was little difference among the different genres they were producing. A laboratory report closely resembled a proposal. They were not learning to differentiate among the various engineering genres and their social actions (C. Miller 1994b).

Clearly, instructors are restricted in the experiences they can provide students about real-world engineering, particularly early on in their education. At the same time that students are learning the content and conventions of the discipline, they are also applying them to engineering problems. In short, they are often placed in a dual position, that of student and practitioner (Burnett 1993; MacKay 1997). To expedite the process of transforming engineering students into engineers, however, instructors designing courses can be aware of some of the issues that emerged in this study.

Integrating the various symbol systems throughout the introductory design course might orient students differently toward written prose. I am not suggesting that mechanical engineering courses emphasize the verbal over the visual, but I am suggesting that the verbal is every bit as integral as the visual. While a mechanical engineer's primary way of knowing is often through the visual, his or her communicating of that information to others is often through the written. . . . and in the form of an argument. Visuals, like numbers, do not speak for themselves. They must be interpreted and made persuasive to others. Throughout the design process, students do this through email, memos, progress reports, and, most explicitly, through proposals. While the theoretical and the mathematical might be foregrounded in their reading of the visual, engineers embed their beliefs about what constitutes a technically sound device within written prose that is meant to be persuasive. From a rhetorical point of view, the theoretical, mathematical, visual, and verbal are interdependent. Each supports the meaning of the other and in the case of engineering contributes to the overall process of the designing of a device.

If the rhetorical nature of engineering design were made explicit to students, they might begin to understand the mutual, ongoing relationships of the multiple symbol systems that come into play as a design evolves and is refined. Rather than viewing writing as an activity ancillary to the design process, they might see its centrality. As shown in this introductory course, the visual must speak for itself when the audience is the machinist and the process has become a product, ready to be built. But when the audience is not the machinist, and the concept is in progress, the design must be explicated so that aspects of it can be approved, contested, and, ultimately, transformed. Engineering is as much about arguing for the technically sound, marketable product as it is about building it.

NOTE

The data for this research was conducted prior to the establishment of the center and, therefore, should be read similarly to chapter 2—as a baseline for visual understandings.

7
I SEE WHAT YOU MEAN
Mechanical Engineering Students' Use of Visuals in a Research Paper Assignment

Sarah A. Bell

The use of graphic elements within the technical writing of engineers is integral to conveying their messages. As one popular textbook enthuses, "Nontextual material . . . helps present your information more effectively and gives a polished, professional look to your work" (Beer and McMurrey 2009, 151). However, these authors provide only ten pages of instruction in their entire text, saying little about what constitutes a "professional" look and even less about what makes graphical presentation "effective." Instead, they suggest students can rely on the "increasing power and ease of use of graphics software applications" that enables students to "create or adapt graphics without having to be graphics professionals" (151). This lack of instruction is typical. In her review of the twelve most popular technical communication textbooks, Joanna Wolfe (2009) found they consistently "ignore or flout the research and advice of scholars in visual design" (360) and tend to focus on the simplest and least effective methods of data visualization rather than those actually used in professional engineering discourse (363). Additionally, Poe, Lerner, and Craig (2010) found that their students at MIT were often "at a loss of what to do with the data tables and figures they have produced" (115) and likely to just dump raw data tables into reports and research papers, assuming that these graphical elements will support claims without further explanation. Technical communication textbooks all agree about the importance of visual communication in engineering discourse, but they may not meet the mark for instruction about how to create and use the types of graphics that engineers consistently rely on: tables, and technical illustrations and diagrams (Wolfe 2009, 363).

With these research findings in mind, I was interested in developing some curricular interventions for the first- and second-year mechanical engineering courses in which I was the embedded technical writing

DOI: 10.7330/9781607328032.c007

consultant. The engineering curriculum for this introductory sequence of four core courses had been significantly revised a few years earlier and was focused around the completion of a design project in each course. Writing assignments consisted of technical memos, proposals, and final reports all related to the design projects. A few graphical elements had been written into the assignments, including design sketches, team logo designs, and tables of data specific to each project, but instruction in creating these was minimal. In the first-year courses, students were given specific templates to use for their writing assignments, and some templates even included boilerplate text and preformatted tables. In the second year, students did not use templates but had assignment sheets that directed them to write specific sections in each assignment and not to deviate from the outlines provided. Evaluations of the students' writing were based on a rubric with four categories: defining a purpose, adapting to the audience, organizing and developing ideas, and clarity. Generally speaking, sketches were evaluated as either pass or fail, and feedback on data tables and plots was limited to the formatting of captions and in-text references. Students also received technical grades from engineering graduate assistants, but these were often given without much feedback. As students in the second-year courses gained a bit more freedom and responsibility to determine the content to incorporate into their assignments, including the most appropriate graphical elements, I noticed a great deal of variability that suggested more instruction would be appropriate. Some students took a "less is more" approach to including graphical elements, and some took an "everything including the kitchen sink" approach, by, for example, including multiple photos of design iterations when a final, well-labeled diagram might have been a more effective choice. It was clear to me that, as with all communication instruction, students needed more than just formatting guidelines; they needed to gain a rhetorical understanding of the role of graphical elements in their technical writing.

In order to establish the most significant gaps in students' understanding, I conducted a rhetorical analysis of a set of research papers written during the third course in the four-core course sequence, a course on thermodynamics. More than sixty students were assigned the writing of a research paper on a topic of their choice related to sustainability. Although the assignment sheet did not prompt them to do so, more than one-third of students (23, $n = 64$) included at least one graphical element in their papers. This afforded me the opportunity to investigate how and why students included graphics within a naturally occurring "control" group where there was no instruction or incentive to do so.

I hoped to identify both effective and ineffective rhetorical strategies students used by looking at three things: the types of graphics used, the placement of graphics within the organization of the text, and textual metadiscourse about the graphics. These categories were identified through a review of the literature as being indicative of effective practice for engineering genres (Donnell 2005; Henderson 1991; Kumpf 2000; Winsor 1996). A very brief overview of this literature follows.

RHETORICAL ROLES OF GRAPHICAL REPRESENTATION IN ENGINEERING GENRES

It is common for historians, philosophers, and sociologists of science to study visualization practices, though Michael Lynch, who provides a succinct review of the research, stresses that these studies are mostly about the "practices that compose and establish the evidential significance" of images (Lynch 2006, 26), rather than images themselves. He finds, "Rather than being a discrete, well-bounded aspect of science, visualization is intertwined with observational and experimental practices, literary representations, and methods for disseminating scientific results" (27). Various forms of terminology—including "depicting," "imaging," "illustrating," "indexing," "demonstrating," and so forth—"alert us to diverse practices that do not fit easily under a single conception of observation or representation" (27). That is, graphical elements serve multiple rhetorical purposes between texts and between contexts. This suggests both a complex terrain for students to learn to navigate in their own communication, and the need for discipline-specific studies of rhetorical practice.

In the case of engineering, John Robinson (1998) summarizes the typical rhetorical context as "explaining why a particular solution to a problem is the best," where "best" can mean "proved optimal through mathematical analysis or other deductive reasoning," or, when this is impossible, "the solution which is judged the most suitable tradeoff" (228). The evidence justifying "best" can come in many forms, but is often dependent on visual explanation. Jeffrey Donnell (2005) states that graphics are integral, even privileged in the rhetorical process by which technical evidence in engineering is created. He observes: "To demonstrate that designs are safe . . . engineers integrate drawings, diagrams, and equations using only a thin thread of text. In data presentations, words may play such small roles as citing figures, explaining labeled features of drawings or plots, defining variables in equations, or establishing the order in which the images of a set are to be examined"

(240). Providing support for this claim, Wolfe (2009) counted the graphic elements in both professional and academic engineering reports and found an average of one graphic element per every 1.9 pages in professional reports and an average of 1.1 graphical elements *per page* in academic reports. It is clear from these studies that graphics are not merely adornments that enhance alphabetic text in the writing of engineers.

Although the relationship between visuals and argument has been studied in professional scientific communication (for example, Gross, Harmon, and Reidy 2002; Pauwels 2006), it has not been widely studied as a strategy in the writing of engineers or of engineering students. Therefore, some generalizations must be drawn from more general principles for visual communication. Engineering discourse can be said to primarily use what Edward Tufte (1997) calls visual explanations, or "pictures of verbs, the representation of mechanism and motion, of process and dynamics, of causes and effects, of explanation and narrative" (10). Winsor (1996) agrees, reminding us that "engineering is knowledge work" (5) that (often) creates knowledge about objects to be built, and it follows that illustrations of these objects are a key visual element of engineering discourse. As such, it is part of a tradition no less illustrious than that started by Leonardo da Vinci, who made full use of emerging techniques for technical illustration including perspective, exploded views, geometrical schematizations, chiaroscuro, and even event chains suggestive of storyboarding in order to examine mechanical objects in depth. But the analysis of statistical data, especially in modeling and testing, is also a primary competency of engineers, and requires an additional communication skill set: the ability to visualize data in order to support design decisions.

Illustrating systematic relations between measured quantities is essential to establishing warrants for cause and effect in the arguments of engineers, but Tufte (1997) demonstrates how common it is for visual explanations to lack the kinds of quantifying markers that allow them to answer the questions of *how many, how often, how much,* and *at what rate.* Tufte's many examples prove that both accuracy and effectiveness are achieved through design, not just labeling and captioning. Axis manipulation is one strategy through which aspects of data can be either highlighted or hidden. A truncated Y-axis, for example, can make small differences appear dramatic. Even when not intended to be misleading, design strategies can make a difference in how patterns in data are understood.

While technical communication textbooks typically suffice for providing definitions of the types of graphics that may provide the kinds of visual explanations Tufte discusses, they do little to explain how to design these graphics so that they do communicate these complex mechanical and/or statistical relationships. As Poe, Lerner, and Craig found (2010): "A researcher goes into a project with a clear purpose and a good sense of what is expected but not necessarily what is going to be interesting about the results . . . Selecting the interesting data from the outputs, the researcher returns repeatedly to the data for further analysis in finally making his or her final case with the visual evidence" (145). Tufte's own maxim is pointed: "For information displays, design reasoning must correspond to scientific reasoning" (53). Engineering students are not well served by the advice to just use the default visualization tools of their engineering and word-processing software. Rhetorically effective visualization requires an awareness of one's purpose, an acknowledgment of the audience's needs, and a full toolbox of visualization techniques.

ANALYSIS OF REPORTS

The number of graphic elements from all reports was tallied. Of 64 submitted papers, 23 (36%) contained at least one graphical element within the text of the report. There were 70 total graphics among the 23 papers. The mean number of graphics was 3.04 with a mode of 2 and a range of 8. Two of the total graphical elements were included on title pages (a university logo and a photograph) and were removed from the analysis, leaving a set of 68 graphical elements for study. Many of the reports analyzed also included what Charles Kostelnick (1996) has called "supra-textual" graphical elements, including report covers; the styling of headings, headers, and footers; and so forth, but I excluded these from the analysis. Although these elements may contribute to the overall authority and professional appearance of a document, I did not find that they were otherwise significant to the informational purposes of the reports in this study. Indeed, most were recognizable as page layout themes from Microsoft Word®.

Interestingly, the use of graphical elements appeared not to have had an unduly positive impact on the grade a paper received. The average grade for papers that did not include graphical elements was 83.04 percent, and the average grade for papers that did include graphics was 84.67 percent. Grades from A–(92.5 percent) to C (75 percent) were recorded between both sets of papers. The average grade among all sixty-four papers was 83.63 percent (B–).

The sixty-eight graphical elements were categorized according to my own schema. Between three technical communication textbooks written specifically for engineering students, I found three different sets of definitions for graphical elements common to engineering genres. Leo Finkelstein (2004) included six categories: equations and formulas, diagrams, graphs, schematics, tables, and photographs. David Beer and David McMurrey (2009) included three: tables, charts and graphs, and illustrations. Sheryl Sorby and William Bulleit (2006) took a more rhetorical approach, dividing graphics into three categories according to purpose: representations of data, charting structure, and portrayal of realism. I found all of these schemes problematic. Sorby and Bulleit's approach, though rhetorical, does not account for potential overlaps in purpose between categories, and the rhetorical categories themselves are limited (what is the purpose of "representing data," for example). I felt that given the importance of both diagrams and schematics in the writing of mechanical engineers, Beer and McMurrey's category of "illustrations" was too broad. Finkelstein's categories seemed to come closest to those representative of the writing of engineers, but his definitions of diagrams and schematics were not precise enough (both included steps in a process), and his inclusion of equations and formulas seemed inaccurate. As many other textbooks do, I consider the communication of calculations to be a separate body of rhetorical experience from graphical communication. Therefore, I defined the following categories based on research about engineering genres (cf. Wolfe 2009):

- diagrams—drawings that show the components of a mechanism, or the relationship among parts of a system;
- schematics—drawings that show the procedures involved in a process or movement through a system;
- illustrations—realistic representations of objects or events (e.g., photographs, artistic renderings, unlabeled models);
- data tables—tabular arrangements of alphanumeric and/or other data (in columns and rows);
- data graphics—visual representations of relations among sets of quantitative data (e.g., charts, graphs, plots).

These categories accounted for all the graphical elements analyzed in the reports as follows:

- illustrations: 19
    (including 11 artist's renderings, 7 photographs, and 1 SolidWorks model);
- diagrams: 16;
- schematics: 11;
- data tables: 11;

- data graphics: 11
    (including 7 graphs, 2 charts, 1 map, and 1 scatterplot).

Table 7.1 includes the breakdown of graphical elements by category for all twenty-three reports in the study corpus. Although no analysis was made between grades and the types or numbers of graphical elements, the papers are listed in descending order with the highest grades first. Papers were also categorized as either making a claim (C) or being exploratory (E). The assignment directions allowed students to either explore a topic related to sustainability or to advocate for a specific technology or policy. A typical thesis statement for a report making a claim was of the format "One solution for problem X is technology Y." A typical thesis statement for an exploratory report was of the format "I will investigate whether or not technology X can solve problem Y." Twelve papers made claims, and eleven were exploratory. Fifty-six percent of the papers included at least one data table or data graphic. Illustrations of data are often associated with evidence for backing up a claim (cf. Poe, Lerner, and Craig 2010), but there was only a small difference in the use of data tables and graphics between the percentage of papers that made a claim (66%) and those that were exploratory (45%). This is likely due to the synthetic nature of the research assignment; that is, writers were not reporting their own research but, rather, synthesizing the research of others.

Two additional patterns are immediately apparent. First, almost all writers captioned figures and tables, which is not surprising given that this was one piece of instruction students had received throughout their previous coursework. It should be noted that the formatting and amount and kind of information included in captions varied significantly, suggesting that the standards of a disciplinary style guide might help clarify expectations for student writers. Additionally, two-thirds of writers included some in-text metadiscourse about the graphics in their reports. In reviewing the literature on text/image collaboration, Hagan asserts that images are "looked at" in terms of viewers' (readers') own history and interests, concluding that "differences in looking can lead to important differences in interpretation" (2007, 50). Put another way, images invite imagined text in the same way that text invites imagined images (i.e., "in the mind's eye"), calling into question the possibility of a unified, or even a cultural "image grammar" (Kress and van Leeuwen 1996), though certain disciplinary conventions may have "grip" (Kostelnick and Hassett 2003). Most technical communication textbooks agree that in order to effectively further the goals of the text, graphical elements need metadiscourse, that is, text that helps readers assign meaning to the graphic, as images are prone to significant differences in emphasis and

Table 7.1. The Types and Total Number of Graphical Elements in All 23 Reports Included in the Study Corpus

Paper # / Type	Diagrams	Schematics	Illustrations	Data Tables	Data Graphics	Total
1* / C	2		3			5
2 / E				1		1
3 / C		1			1	2
4 / C	1		2		1	4
5* / C	5		1	1	1	8
6 / E	1				1	2
7* / C	3		1			4
8 / E	1					1
9 / E	1					1
10 / C		1	3	2		6
11* / C		2	1			3
12*† / C		1				1
13 / E		2				2
14* / E	1		1	2	3	7
15† / E				2	1	3
16 / C			4		1	5
17 / C				1	1	2
18† / E					1	1
19 / E			2			2
20 / E	1	1				2
21 / E		2				2
22* / C		1		1		2
23 / C			1	1		2

*Paper did not include in-text metadiscourse about graphical elements.
†Paper did not include figure/table captions.

interpretation or even misunderstanding by readers. In the study corpus, the amount of metadiscourse per graphical element varied from none to an entire paragraph. The amount and sophistication of metadiscourse varied by writer and seemed unrelated to the type of graphical element being used. A typical amount of metadiscourse was a sentence or two referencing the placement of the graphical element; for example, "This reaction causes water, air and heat to flow out of the right side as shown in the diagram below. This process continually ejects heat, or the energy that will be converted to a practical medium for a variety of uses." In some cases writers did draw readers' attention to specific parts of graphics or interpreted graphics for readers. In one particularly successful example, the writer used metadiscourse to compare data tables from two different studies that were placed on different pages of the text.

The second pattern is that two-thirds of writers included at least one diagram or schematic in their reports. Although there were more total illustrations used (as described above), it is noteworthy that the most *common* graphical choice (70% of papers) was a diagram or schematic possibly the most widely used type of graphic in mechanical engineering. By way of comparison, 57 percent of papers included a data table or data graphic, and 43 percent included an illustration. Since the vast majority of reports in the study corpus explored or advocated for the continuing research and development of specific technologies, it is appropriate that writers would include graphical elements that describe the mechanics or processes involved in the technology being discussed. The remaining discussion in this study will analyze writers' rhetorical use of these three groupings within the study corpus: diagrams and schematics, data tables and data graphics, and illustrations.

Diagrams and Schematics

Diagrams and schematics depict processes via abstractions of system flows, and labeling that stresses the working of components within a system. Whether making a claim or not, writers typically included a diagram or schematic of the working process of the primary technology being discussed in the report, for example, a dry steam power plant, a solar cell, a wave terminator, a fuel cell, a wastewater treatment plant, or a septic tank. As such, the diagrams and schematics were often included near the beginning of the body of the paper as an overview to the topic. As one of the key graphical types used in engineering design, it makes intuitive sense that even without instruction, students have accurate ideas about the use of diagrams and schematics, picked up from their information environment. This may also be indicative of engineering students' "information ground" (Pettigrew 1999).

While writers typically used diagrams and schematics that accurately matched the device or system being described in the report, they also typically assumed these diagrams and schematics were self-explanatory, including little metadiscourse about them, leaving the reader without much interpretative focus (Mathison, previous chapter). Additionally, because the diagrams and schematics were being lifted from other sources rather than created by the students themselves, the graphics often included much more information than was discussed within the text of the students' reports. At times, this incongruence between graphic and text raised questions about whether the writer's discussion was accurate and thorough. At other times, the writer's terminology did not match the labeling of the graphic, causing confusion for readers. For example, one student wrote, "A terminator is a device that can absorbs, transmits and reflects all the energy in a wave [*sic*]," whereas the actual diagram used more technical language to identify wave rays, wave crests, parabolic walls, turbines, and so on. In that writer's next example, he has used more textual description of a structure "oriented roughly parallel to the direction of wave propagation . . . [and] motion is used to pressurize an internal hydraulic fluid. This pressurized fluid turns a turbine that is coupled to a generator," but the accompanying diagram only labels the pontoons of the device. While this can perhaps be seen as the writer's attempt to fill in the gaps for diagrams lacking labeling, while letting labeling speak for itself in others, neither of these strategies is as effective as a more parallel interplay between text and diagram would be.

Ideally, students' written texts would support the interpretation of and focus on diagrams and schematics. An example of a writer who did an effective job of utilizing a schematic in his text introduced the graphic with the metadiscursive statement "As shown in the figure below, the air is pushed upwards and out through a valve," which is indeed followed by an abstracted figure of an oscillating water column device with the air flows, air turbine, and valve clearly labeled. The writer follows the figure by highlighting what he wants readers to focus on when looking at it: "This design is of particular interest due to the development of a two-way turbine that is able to constantly spin in the same direction, regardless of the direction of air. In other words, as the air retracts back into the column as the fluids level lower the turbine will be able to rotate the shaft in the same direction, creating more energy efficiency [*sic*]." Although not as clearly stated as could be, the text is supported by arrows and labels on the figure that also make this point; consequently, both figure and text direct readers to the main concepts this writer wants to make clear.

Data Tables and Data Graphics

In their case studies of student writers at MIT, Poe, Lerner, and Craig (2010) bluntly acknowledge that "in scientific research articles, visual representations of data are the work horses of argument" (115), comprising, on average, 26 percent of the surface area of contemporary publications, broken down as 18 percent figures and 8 percent tables (Gross, Harmon, and Reidy 2002, 200). Engineers have been found to follow similar use patterns (Wolfe 2009). It is reasonable to assume that writers in this study understood data visualization to be a key characteristic of engineering genres, as 56 percent of reports included at least one; however, of all the types of graphical elements used, writers proved least adept at using data tables and graphics to accurately support their own positions. Of 22 total data visualizations (11 tables and 11 other data graphics) 19 (86%) were found to be ineffective because

- they were missing key aspects of the data needed to understand or make use of it in the context of the writer's discussion (5 visualizations);
- they were misread (or misrepresented) by the writer (3 visualizations);
- they contained too much extra data irrelevant to the writer's discussion (3 visualizations);
- they were unreadable (3 visualizations);
- they didn't present enough data, or text would have been more efficient (3 visualizations); or
- there was no context provided to understand the data (2 visualizations).

The most common problems, missing or misreading data, can also be the most significant as they leave readers most unprepared to accurately interpret the data being displayed. In one example, the writer provided a scatterplot (imprecisely described as a "chart" in the report) but failed to explain what was plotted. Another writer claimed that the table provided proved that high-speed rail is the "clear winner" among a spectrum of passenger vehicles when compared according to emissions per passenger mile; however, the table clearly indicated that both bus and conventional rail systems have lower emissions per passenger mile than high-speed rail, a fact undermining the writer's argument.

Another typical problem was inserting a data graphic that didn't match up with the writer's text. For example, a graph in one report was supposed to demonstrate that repurposing cell phones is more efficient than recycling them, but the graph only provided data about the declining mass and value of recoverable gold from cell phones and did not provide enough data to support the larger argument. A graph in another paper showed a rise in the index for food, oil, and other commodities across the years 1992–2008. The writer discussed the impact of commodity prices on developing countries, but didn't provide enough information about the graph to let readers know if it was measuring commodities indexes across the world, or in the United States, or within some other political or economic boundary. The graph further undermined the writer's argument about food prices becoming increasingly burdensome because the graph showed food prices remaining fairly steady while oil prices skyrocket. While a relationship could be drawn, the text fails to do this. Another writer incorrectly represented a table *projecting* sales of electric vehicles in the United States over the next eight years as *proof* that the number of electric vehicles "will" rise. In a complex table in this same paper, the student claimed that an increase in battery-charging infrastructure was shown to change the price of electric vehicles when, in fact, the table shows that even with a ten-year battery amortization, the federal government would have to subsidize the price of electric cars by half in order to equal the per-mile cost of gas-powered vehicles. Any information about an increase in charging infrastructure had been left out of this version of the table.

A few illustrations of data could have been made much more effective if the writer had taken data from the sources and created his or her own graphic, rather than copying data graphics directly from the outside source. For example, one writer used two separate tables from two different sources to compare different greenhouse gas emissions from two different technologies. Because the tables didn't match in design or order of information, a lot of "noise" was created for readers trying to see the comparison being made. For example, one table was formatted horizontally and the other vertically. One provided emissions in pounds per megawatt hour, while the other did not specify units. While not easily translated, the comparison does help to further the writer's argument. However, it would not have been difficult to design a single table to organize the comparison effectively. In an effective example, the writer explained the material properties of four different plastics within a paragraph in the text and followed this up with a three-dimensional bar graph comparing those same plastics. Since the text and the bar graph matched completely, the bar graph served to help clarify the textual explanation by showing it in visual form. As many sources point out, though (cf. Tufte, Wolfe), bar graphs are typically an overly simplistic data visualization not suitable for engineering discourse, and three-dimensional bar graphs, in particular, create what Tufte (1997) derisively labels "chart junk" that gets in the way of a reader's ability to understand the data being presented.

Illustrations

There is little research available about the use of pictorial illustrations in engineering genres. Unlike for diagrams/schematics and data tables/graphics, it is not known how often pictures occur in engineering publications. In his textbook, Finkelstein focuses specifically on photographs, telling students that photos are a "high-bandwidth" tool that can provide readers with a lot of information quickly, but warning that pictures also run the risk of being irrelevant, or distracting when poorly reproduced. Beyond photographs, I defined illustrations as nonlabeled, realistic representations of objects for the purpose of this analysis because many students, writing about future technologies, used artist's renditions where photographs were unavailable. My analysis found two effective uses of pictorial illustrations, as well as numerous examples of extra pictures being used merely to decorate the text.

The most successful uses of pictorials occurred when pictures were in *parallel* to textual discussions of emerging technologies discussed in the text. Parallelism is a strategy that Jeanne Fahnestock has identified as persuasive, except that she defines visual parallelism specifically as "the juxtaposition of three or more images in either a vertical or horizontal row or in an array of both" (2003, 142). Fahnestock finds such arrangement can constitute an argument. Although I am diluting her notion of parallelism a bit in that no image arrays were found in this corpus, I did find that, as an organizational strategy (rather than an argumentative one), the use of images in parallel to the organization of the text was effective. Fahnestock is drawing a much more specific analogy between grammatical parallelism and visual parallelism than I am. Susan Hagan identifies the existence of "perceptual ties" between images and text when they include shared location, similar location, alignment, and overlap (2007, 52). I am calling such perceptual ties "organizational parallelism."

A representative example occurred in a report on technologies to harness tidal power. The paper included a section with the heading "Types of Underwater Turbines" that discussed three types of turbines: underwater kites, shrouded turbines, and hydrofoils. Each description included an illustration of the turbine type (though since the pictures are each borrowed from different sources, they are not consistent in style, size, clarity, and so forth). Fahnestock also specifically says that for visual parallelism to occur, images must be visually parallel—the same color, background, viewing angle, and so on. Again, I am taking liberties with the notion of "parallelism." Two other papers on tidal and wave energy used similar strategies in which discussions of the technologies were organized by the different types, with a picture of each included for clarity.

CONCLUSION: PEDAGOGICAL INTERVENTIONS

There appears to be considerable room to develop pedagogical interventions to facilitate students' more effective use of graphical elements. "Visualization," or even "graphics," is terminology that belies the spectrum of activities and competencies involved in making anything visible in science and engineering. As such, "it" can't be developed if solely relegated to a "visual literacy" or "graphic design" unit within a technical communications course, though that is certainly one component of a comprehensive pedagogical approach. Visualization develops over the course of any research endeavor, and it should be acknowledged that "in learning how to select, shape, and present data, students learn the ways that professional scientists make arguments with quantitative evidence, and they also are introduced to the constructed nature of the scientific research process" and that "what ends up being presented as visual evidence in the final published research article is a reorganization of raw data in support of particular claims" (Poe, Lerner, and Craig 2010, 114). Based on the successes and failures identified in this study, the following pedagogical interventions might be pursued in an attempt to bring students' abilities to effectively use graphics in line with disciplinary practices.

First, create collaboration and assign apprenticeship genres.

Sociology of science tells us that the lone genius is not the status quo for STEM disciplines. Problems are more typically addressed and solved in groups. As Winsor points out, though, typical rhetorical understandings of audience as unidirectional do not accurately represent the "interactive, recursive relationship" that working engineers often have with their readers (1996, 4). In many ways, the assignment on which this exploratory study is based does not represent the "real world" of writing as an engineer. The other writing assignments these students accomplish in the course of their two-year core curriculum are more representative of apprenticeship genres and include memos, lab reports, design reports, and proposals, all collaboratively constructed. Simultaneously, students are being trained in some freehand sketching, SolidWorks modeling, free body diagramming, design brainstorming, and some data formatting. We could do more to further integrate interplay between these developing visual abilities and the written texts students are asked to produce. For example, the writing instruction can include lessons and feedback on metadiscourse within lab reports about student-generated graphical elements. While most documentation has to be kept from other students in these courses because it describes work toward a design competition, students could still be provided with better models of visual/alphabetic/numerical texts as well as increased feedback from both engineering and writing faculty. However, scaling this feedback to the large numbers of students (150+) and assignments in these courses is a constant challenge.

Second, teach information design along with statistical literacy.

If anything is to be learned from the work of Tufte, it is that effective and ethical graphical information is the art of making *data* visible. Before this can be accomplished, students need to be able to create, manipulate, select, and interpret data. Reporting and visualizing data need to be learned along with these other skills. As noted previously, Poe, Lerner, and Craig found that students often "dumped" data into their reports without analysis (Poe, Lerner, and Craig 2010, 115). This study found that students also "dumped" data into their research papers, often when that data did not support, or even undermined the text. A pedagogical approach which only models the "correct" presentation of standard table and chart formats fails to recognize the cognitive processing and experiential learning that precede the ability to make appropriate formatting decisions. Poe, Lerner, and Craig's case studies support the conclusions that experience with different genres helps build understanding in how to use and present data (they advocate storyboarding as an invention and drafting exercise) and that authentic, professional faculty feedback is necessary to guide students in learning to use data to make arguments. A simple addition to my curriculum could be incorporating Tufte's (1997) famous critique of the *Challenger* disaster information graphics as students are beginning to collect their own data in an effort to teach them how much presentation of data can matter. The findings of this study, though preliminary, clearly demonstrate that the graphical representation of data is a significant stumbling block for students.

Third, integrate instruction and practice in both alphabetic and visual communication throughout the engineering design process.

My purpose is not to (re)privilege the visual over the alphabetic. In her study of a first-year premechanical engineering design course, Mathison, in the previous chapter, found that students often believed the role of writing followed the design process and was only used to describe or present a finished product. From my perspective, the aim of our integrated communications curriculum is to provide students with a complete toolbox from which to draw, that is, the ability to draw from *and combine* the affordances of mathematical, computational, visual, verbal, interpersonal, *and* written literacies throughout the sense-making, problem-solving, and design processes (see Read and Mathison, chapter 4 in this volume). I often show slides of pages from da Vinci notebooks to students in an attempt to keep reinforcing the power of combining all of these ways of knowing across the spectrum of their activities as engineers. Writing is not peripheral or an addendum to the design process, but integral. Likewise, the visual is not merely addendum to "final" documentation. Furthermore, the suggestion of this study is that students need assistance developing their ability to provide metadiscourse to support the use of graphical elements of all types. In this, their engineering faculty and their communication faculty can be key in providing modeling, feedback, and guidance in developing rhetorical skills in the interplay between the visual, the numeric, and the alphabetic.

Even though half of university engineering programs in the United States require a course in technical communication (Reave 2004), research suggests their effectiveness might be questionable if students are primarily gaining skills in formatting (Ford 2004) rather than higher-order rhetorical concerns such as purpose and audience. There is evidence that tight integration between communication and the engineering curriculum is a more successful model in this regard (Kryder 1999; Poe, Lerner, and Craig 2010; Sageev and Romanowski 2001). The findings of this study support the idea that students need this integrated approach in order to develop rhetorically sophisticated strategies for merging visual and alphabetic elements in their texts and that they need help developing visual and writing competencies from both their engineering and communication faculties.

8
IDEOLOGIES OF GENDER
Culture Clash between the Disciplines

April A. Kedrowicz and Julie L. Taylor

Engineering students typically have developed expectations for a preferred teaching style and assumptions about what their educational experience should be like. When their learning environment is "disrupted" due to the presence of outsiders teaching non-STEM (but STEM-relevant) content as is often the case with WID and CID initiatives (in this chapter), their identities can be challenged, creating a potential barrier to successful interdisciplinary collaboration.

Underlying students' expectations and assumptions are deeply held disciplinary ideological beliefs that operate at an unconscious level, yet inform their assumptions and expectations. Ideology informs values, norms, practices, *and* gender dynamics (e.g., Sullivan and Kedrowicz 2012), particularly when Humanities and STEM disciplines collaborate. The presentation of information, associated discourse, and presence of particular bodies (men versus women) can serve to mark a specific field—STEM disciplines are marked as masculine and Humanities as feminine (McCloskey 1985). Moreover, the gendering of disciplines can serve to position them hierarchically in that masculine disciplines and ways of knowing tend to be privileged over their feminine counterparts (e.g., medicine vs. nursing). This added layer of gender can further problematize an already challenging interdisciplinary WID/CID collaboration.

In this chapter, we examine how engineering students' expectations for educational experiences inform their reactions to "feminine" communication instruction and instructors in the classroom. Specifically, we highlight how students' reactions to this collaboration stem from ethnocentrism and stereotyping that can impede educational goals. We offer recommendations for practitioners so they can capitalize on strengths and mitigate challenges characterizing interdisciplinary collaboration.

DOI: 10.7330/9781607328032.c008

DISCIPLINES AS GENDERED

Disciplinary cultures are characterized as gendered, which can be seen through their privileged discursive structures. That is, use of specific discourses can both illuminate and (re)produce organizing cultural preference. Katie Sullivan and April Kedrowicz (2012) explained, "Academic disciplines constitute cultures with their own sets of rules, norms, and gendered expectations" that are present within their discourses (597). The difference between disciplinary cultures is perhaps most noticeable when academic cultures collaborate. In other words, ideological concerns are likely to surface and create tension stemming from unique beliefs that clash. More specifically, the attributes of gender dynamics—which are ever present in the engineering classroom—stemming from disciplinary norms, makes for an ideal study of disciplinary cultures through examining the organizing discourse (for a history of gender formation in the technical sciences, see Oldenziel 1999).

DISCURSIVE ORGANIZING: MASCULINE AND FEMININE

Organizational communication scholars have argued that discourse is the key to understanding organizing (e.g., Fairhurst and Putnam 2014; Hardy 2001). Furthermore, discourse becomes a central focus for scholars to examine preference, privilege, or difference within an organizational setting. To examine discursive organizing, we take the position that discourse represents both macro- and micropractices, which inform rules and structures within an institutional setting (Mills 2004; Weedon 1997). Microdiscourses refer to the actual words and language choices used, whereas macrodiscourses are informed by and inform larger ideological and cultural preferences. It's important to recognize both micro- and macrodiscourses as symbiotically related, each as an organizing element of culture (Fairhurst and Putnam 2014). Moreover, language (i.e., microdiscourse) used can both highlight and (re)produce the gendered culture, and the culture, in turn, informs preferred language choices.

Language is the way that we construct and communicate our realities (Weedon 1997). More specifically, language materializes social structures, preferences, and contexts, and in essence, contributes to the (re)defining of the social context in which they are a part (Macdonell 1986). As a result, organizing discourse is a way of understanding the ideological power functions within social structures and institutions, because people traditionally use discourse that is supported by the larger structures in order to be productive or accepted within the ideological constraints of the context (Mills 2004). For instance, currently engineering education resides in a masculinist discourse, reflected in its language and culture. Specifically, in engineering the generic "he" is generally used for examples, and masculine styles of communication are privileged over other forms of communication, regardless of actual student demographics. This facilitates development of a gendered engineering identity through the experience of being "male."

Traditional discourse scholars contend that gender is a construct because of language (e.g., Mills 2004; Weedon 1997). Therefore, it is through language that we can understand the social construction and subsequently social organizing of gender. Important to note in this study is the focus on gendered patterns in communication. That is, masculine communication is traditionally characterized as objective, powerful, and inexpressive (Wood 2013). On the other hand feminine communication is characterized as subjective, less powerful, and "erratically" expressive (Wood 2013). Focusing on the characteristics of communication separates gender from sex, such that gender is a performance and sex is biologically determined (Wood 2013). Therefore, organizing is determined though communication (i.e., language norms) not necessarily biological sex.

Former studies within engineering settings suggest that masculine styles of communicating, and thus organizing, are the preference (e.g., Faulkner 2000; National Research Council 2001). Further, in engineering, masculinity is accompanied by an inherent level of knowledge and power that is not typically associated with its female counterpart (Hacker 1989). This is exercised even in the classroom, where students ascribe power and knowledge to their male instructors and expect female instructors to be nice, caring, and understanding (Bachen, McLoughlin, and Garcia 2009). In fact, female instructors sometimes encounter a double bind with respect to their looks and intelligence—the more attractive a female is, the less intelligent she is assumed to be (Tonso 2007). Becky Atkinson's (2008) research on female instructors and power dynamics in the classroom shows that when female instructors challenge power among male students, they often sexualize her in an attempt to reduce any legitimate power she may have.

Within the specific context of an engineering classroom, Sullivan and Kedrowicz (2012) examined the gendered consequences of a female body teaching communication (feminine discipline) in an engineering classroom (masculine discipline). They found that students not only resisted what they define as a "soft" science (oral and written communication), but also devalued the female embodiment of communication. More specifically, students often made comments related to their instructors' bodies, rather than to their competence and skills. Students claimed that the information they were receiving was not technical; it was common sense, and they resisted what they deemed to be marginal information, though engineering is never just "technical" (Faulkner 2007, 336).[1] In the end, Sullivan and Kedrowicz (2012) uncovered a power dynamic in relation to gendered bodies in the engineering classroom, privileging the masculine and devaluing the feminine.

Building on these findings, Julie Taylor (2013) examined the negotiation of difference within the engineering classroom. She explained that gender, as represented through the female body, may be initially rejected, but over time and with proof of competence is negotiated and conditionally accepted (in regard to both student and instructor). The moment of conditional acceptance as described in Taylor's study is what is important to this chapter: How do students perform conditional acceptance? In other words, if what engineering students want (hard knowledge) is incongruent with what they get (soft knowledge), how do they respond and to what consequence? Additionally, how do they perceive the deliverer of soft knowledge, in this case communication instructors?

The deeply entrenched masculinist culture of engineering facilitates intercultural miscommunication stemming from lack of knowledge of the other, ethnocentrism, and stereotypes (Lim 2003) even more likely than it would otherwise be in a discipline whose discourse is more representative. And while communication and writing could be the boundary objects (Mathison and Berkland, chapter 1) that bring interdisciplinary faculty together, the same cannot be said for students who are only beginning their enculturation into the field.

CURRENT DISCOURSE IN ENGINEERING EDUCATION

Masculine discourses are constructed through socialization and thus become normative. Godfrey and Parker (2010) identified six dimensions of engineering culture providing a framework for understanding its disciplinary identity—values, beliefs, and assumptions: (1) an engineering way of thinking, (2) an engineering way of doing, (3) being an engineer, (4) acceptance of difference, (5) relationships, and (6) relationship to environment. An engineering way of thinking refers to tangible, quantifiable, measurable reality and a concern with knowledge that is relevant to everyday life. An engineering way of doing is grounded in the disciplinary characteristic of "hard" or difficult work.

Not only do engineering students learn that anything worth doing is perceived to be difficult, but they also come to devalue subjects areas they consider soft and, thus, considered easy. In fact, "soft concerns have the potential to discredit one's identity as a serious professional" (Hacker 1989, 43). This relates to how to "be an engineer" in that engineering students develop an early identification as engineers—high achieving, "can-do" people.

Students recognize the importance of personal relationships to succeed in the program and generally have good working relationships with engineering faculty best characterized as a mentor/mentee relationship. Generally speaking, conflict and dissent are avoided due to their similar pattern of understanding and communication. There is a strong awareness that the goal of engineering education is to educate graduates so they can function as professional engineers; thus, education should be shaped by the needs of the profession, which, until recently, consisted of courses focusing on "hard" skills and *not* effective communication, a "soft" skill (Godfrey and Parker 2010). While ABET now acknowledges the importance of communication skills in the profession (Shuman, Besterfield-Sacre, and McGourty 2005), practice still lags.

Imperative for our future engineers is the acceptance and integration of the "other" into current practices because skillful interdisciplinary collaboration will be paramount for them working in a global society (ABET 2018; Kedrowicz 2004; Lattuca 2001). Much of engineers' industry work will require they collaborate with people from various backgrounds, areas of expertise, gender, ethnicity, and so forth. However, despite the obvious importance of different perspectives, individuals who exhibit ways of thinking, doing, and being that are "outside the norms" might not fit engineering culture and thus, as Godfrey and Parker (2010) contend, are accepted if they can demonstrate an engineering identity, "being one of the guys" (14). Explained a female operations and research engineering graduate, "You have to be tougher when you are around the guys, you feel you have to do better than them to be accepted" (Fouad et al. 2012,17).

Despite engineering faculty members' commitment to students' development of communication competence, and in many cases, their outward endorsement of the importance of communication alongside the expertise of disciplinary content, most students experience some degree of dissonance in the interdisciplinary classroom. Our objective is to understand students' responses to the communication-engineering collaboration to enhance the experience for both in-group and out-group members.

APPROACH TO THE STUDY

The College of Engineering under study is comprised of six different departments (Bioengineering, Chemical, Civil and Environmental, Electrical and Computer, Materials Science, and Mechanical). In order to narrow our data, we examined student feedback from the two more characteristically masculine departments in order to ensure a substantial amount of data for understanding the impact of gender negotiation. Traditionally, mechanical engineering and electrical engineering are more masculine disciplines and consist of mostly male bodies with respect to both teachers and students (Hacker 1989). At this particular university, fewer than 10 percent of the student and faculty populations in the College of Engineering were female at the time of data collection.

We conducted a thematic analysis of students' open-ended responses to course evaluation prompts from classes including freshman through senior students across two departments spanning seven years. This facilitated rich description and interpretation of the data set to identify both semantic and latent themes. Specifically, we coded 16 sets of evaluations from 8 different electrical and computer engineering courses with 7 different instructors (5 females and 2 males), and 26 sets of evaluations from 6 different mechanical engineering courses with 9 different instructors (5 females and 4 males). For both departments, we gathered progressive information from students at every phase of degree achievement—from freshman through senior years. The evaluations that students complete are anonymous so we were unable to track individual students; however, we were able to see patterns in their understanding or appreciation of the "soft" discipline within their "hard" classes. On the other hand, we did look at themes among WID/CID consultants to examine, for example, if a more feminine-style communicator would elicit certain responses over a more masculine-style communicator within the hard discipline. The findings to these themes will be presented in the following section. In addition, pseudonyms have been used for all parties involved with respect to their positions and contributions.

The evaluations, which are individually crafted for each course, are distributed near the end of each semester and contain a combination of open and closed prompts. The open-ended questions are designed to elicit student feedback about the effectiveness of instruction, the quality of communication integration, and perceptions of the instructor. For example, students were asked questions similar to the following: (1) What are two ways you see yourself using the oral communication skills you learned in this class in your academic and professional life? (2) Overall, what is your impression of the integration of communication skills into

Table 8.1. Data Corpus

Discipline	Number of Sets of Evaluations	Number of Courses Evaluated	Number of Different Instructors	Sex of Instructor
Electrical engineering	16	8	7	5 Females 2 Males
Mechanical engineering	26	6	9	5 Females 4 Males
Total	42	14	16	10 Females 6 Males

engineering courses? (3) Give two examples of how the writing instructor's lecture helped you write a design proposal. Students' responses to all open-ended questions on forty-two different sets of course evaluations became the data set for this study. We chose not to combine all course evaluations across departments and courses so that themes among specific instructors and classes could potentially emerge. Table 8.1 illustrates the data corpus.

An inductive, iterative approach guided our data analysis (Miles and Huberman 1994). Both authors coded (Corbin and Strauss 1998) every response to each course evaluation simultaneously. We read each response, noting key words or phrases, and recording initial thoughts and reactions. This allowed us to pay attention to emergent patterns within and across courses and instructors. We then looked for thematic categories among students' responses as they related to instruction and the instructors in oral and written communication. While this was labor intensive, it allowed us to discuss ideas and discrepancies as they arose and engage in an iterative process of data analysis. The next phase of analysis consisted of grouping categories together and making connections between categories using the constant comparative method (Lincoln and Guba 1985; Tracy 2013). We relied on our extensive experience with interdisciplinary collaboration in this setting to guide the formation of impressions grounded in students' expressed understandings, attitudes, and beliefs. Of note, the first author had over ten years of experience working with engineering faculty and students at the time of data analysis. As such, experiences in faculty meetings, observing and teaching classes, facilitating workshops, interacting with students informally, and preparing end-of-year reports and publications impacted and informed the development of this analytical framework. The final phase focused on identifying relationships among the themes, or "explicating the storyline" (Corbin and Strauss 1998) to develop a coherent picture of students' experiences with (1) communication instruction and (2) communication instructors in the engineering classroom. In the next section, we report on our findings of engineering students' responses to communication instruction and instructors by highlighting the interrelationships between prevalent themes.

INSTRUCTIONAL THEMES OF ENGAGING WITH THE FEMININE IN THE CLASSROOM

Three themes emerged when analyzing students' comments regarding instruction: (1) utility of instruction, (2) perceived authenticity of instruction, and (3) resistance to instruction. Combined, the three themes highlight the importance of communicating in engineering yet cast doubt on its relevance when taught by instructors from a feminized discipline; the interplay of utility and authenticity resulted in many evaluative comments that indicated resistance to "out-group" expertise, which was deemed as delivering "soft" rather than "hard" knowledge. While students are socialized to believe that communication is an important component of engineering practice, their beliefs are also embedded within an historically disciplinary masculinist culture, perceived as clashing with those of a feminist disciplinary culture.

Utility

Data revealed that students see significance in the future utility of communication to engineering work. They appear to understand that writing can be a means to attaining success across a number of measures in the workplace. First, students appear to see to the value in learning and practicing their communication skills because they believe these skills will be useful to them in the future, specifically with respect to professional promotion and earnings potential. A sophomore student acknowledged the importance of communication skills: "[Communication will be important] when I become CEO of an engineering firm, giving presentations." A senior student claimed, "Professional presentations will certainly be a part of my career," thus showing that these students acknowledge and respect how and where communication aspects will be relevant in their future careers. More noticeable is their recognition of the connection between better communication skills and career progression. As stated by a freshman student, "They [communication skills] are very important, people with good communication skills move up." Students at all levels of their education seem to appreciate the role of communication to their professional development, perhaps due to early socialization experiences and interactions with engineering faculty who emphasize skill development in this area. When discussing the relevance of communication to engineering, many faculty members point directly to the relationship between effective communication and managerial success.

Interestingly, a few students also tie communication competence to the ability to get engineering projects funded: "[Communication is important for] giving updates and making proposals to get money for projects." Likewise, "Despite the merit of any project, if you can't communicate its value, you will never get funding." Engineering students value what they perceive they will need for future successes. Having early identification of how to be an engineer, through enculturation, engineers understand and communicate the importance of "selling" their ideas in the face of competitive contracts. In short, communication is a means to an end and represents one specific tool in their identity toolbox (Gee 1989) for securing a position and achieving success in the profession. Yet, utility, it seems, is not enough for students to embrace communication instruction and activities in the engineering classroom. Data also revealed that only a specific type, or what students believed was *authentic* communication, has value, and this view of authenticity is rooted in an ethnocentrism about the hierarchical positioning of the engineering discipline as well as stereotypes about communication instructors.

Authenticity

Students' overarching concern centered on their belief that school and, particularly, professional communication assignments and activities are somehow inauthentic or contrived, arguably made more so by the presence of nonengineering instructors. In fact, students repeatedly called for "real" examples that would be used in "real life" or the "real world." For example, one student commented that "[I would like] real world information. We need to know what is out there; maybe some guest lectures/real memos from engineering business as examples." Another said, "I would like to hear someone who has actual engineering experience talk about communication. I feel that our instruction might not be too closely related to real life."

Students lacked trust in the education they were receiving with respect to communication, despite having content experts collaborate with engineers to design the curriculum. One student went so far as to ask for "proof" of the authenticity for what they were learning: "[I would like] proof of use in the business place. Bring in businesses who [*sic*] communicate. Don't use education derivations." This comment could allude to students' mistrust of classroom instruction, perhaps made more prominent due to the presence of communication instructors (disciplinary out-group members) and a perception about their limited skill and knowledge of the engineering workplace. Or, it could also point to students' views on classroom activities as mere "hoops to jump through" on their path to becoming an engineer. Because communication is typically not a skill commensurate with an engineering way of doing things, students desire some sort of tangible proof that what they are learning will in fact be relevant to their future as engineering professionals, or to their life in the "real world." An engineering way of thinking values knowledge that is concrete and authentic, such that students (and faculty) expect content and learning to be contextual, practical, and relevant.

Resistance

Despite the recognition of the importance of communication to engineering, students resisted communication instruction and consultants. Their comments revealed that they thought communication detracted from learning "real" engineering; their perceptions revealed skepticism about the value and authenticity of communication activities and assignments and the credibility of communication instructors.

Much of this resistance appeared to stem from the fact that communication instructors lacked a background in engineering or a closely related technical field, as this comment illustrates: "They didn't have an engineering background, so in my opinion, they weren't entitled to know what they were talking about." Inherent in this comment is the idea that communication consultants—trained in communication theory, pedagogy, and professional competence—are not qualified to teach speaking and writing to engineering students. In this context, engineering expertise trumps communication expertise, even when the subject being taught and assessed is outside the engineering discipline. The discourse reminds us that engineering culture produces and (re)produces the practices of "only" doing and being an engineer. Godfrey and Parker (2010) state, "Differences in gender, race, or prior experiences are tolerated and accepted provided the attributes of 'being an engineer' are met" (15). For example, engineering students' ethnocentric views on the superiority of the discipline of engineering over communication can impede their ability to fully engage with and benefit from communication instruction. Somewhat more troubling is the discursive trend suggesting that engineering undergraduates feel compelled to judge pedagogical competence and content expertise in communication, a field where they are very much outsiders. One student remarked, "In ECE classes, having English [sic] majors edit our papers is a total waste of everyone's time because they don't have the technical knowledge to give effective feedback." Again, we see the underlying assumption that technical knowledge is a requirement—the only requirement—necessary to give feedback on writing. From a cultural perspective, this could relate to students' hierarchical ordering of content over form. Assuming that engineering students, up to this point, have generally been shown that only content matters because of its emphasis in the classroom, they seem to lack the sophisticated understanding of the synergistic interplay of content *and* form and, as evidenced, do not understand the inseparable relationship between the two. We see evidence of stereotyping in these comments; students have misperceptions about the skills and benefits of having communication consultants work with them to improve their writing and speaking competence (i.e., synergy of form and content), perhaps related to their devaluing of the soft, "easy" content of communication.

In fact, some students specifically called out the communication instruction for being "common sense" and expressed this as the specific reason for not relying on feedback and/or calling for more rigorous instruction as these comments suggest: "The instruction on teamwork was basic and seemed like common sense. More depth on the subject would be helpful." "[Provide] more insightful information. A lot of it was common sense." "Most of it is common sense of how to act in front of people." "I didn't use [the feedback] because it was obvious stuff." Students view instruction and feedback on communication as basic knowledge and, presumably, something anyone can do since it is merely "talking to people," a tool they already have in their toolbox because they have been talking and writing for years. It appears that students have embraced engineering culture and ascribe to the idea that anything worth doing is hard, which is decoded as scientific, mathematical, and technical, highlighting Sally Hacker's (1989) reminder that communication, a soft skill taught by disciplinary outsiders, is not part of the serious work of engineering.

A prominent theme surrounding students' resistance, and related to communication as a soft, feminine skill, is the notion of interpretative subjectivity. Students expressed frustration with feedback that was merely "opinion" with no basis in "fact" as these comments illustrate:

"[I would like] actual writing standards instead of personal opinion." "Ken (the consultant) needs to understand that his ideas are just his opinions NOT facts. Especially on the resume."

In fact, students' comments point to direct and overt resistance to feedback from communication instructors: We also found in the comments that students disregarded feedback they received from communication consultants:

"I haven't looked at most of the comments our papers, so I guess I didn't utilize the feedback at all." "I might use [the feedback] to get a good laugh."

Peer feedback, arguably from novice writers / speakers, was viewed as more beneficial than feedback from the communication specialist. In one student's words, "Feedback was subjective, based on the observations of a communication specialist. I found feedback from my peers to be more useful." In other words, specialized knowledge and experience did not translate into useful feedback for this student, perhaps due to the assumption that soft expertise is not as useful or reliable as hard, objective information from someone who embodies an engineering way of thinking and doing, albeit at a novice level. Thus, the desire for, and practice of, feedback is once again informed by the standards set by "hard" ways of knowing. One student specifically referenced the importance quantifiable feedback: "The feedback did not really help. Does not quantify deductions in grade. Rather than 'do this, do that, improve this, improve that,' I would prefer 'you missed 5 points for . . .'" It seems, then, that students wanted soft information translated into hard information.

Along the same lines, students expressed frustration with the subjective (soft) "writing" portion of the grade counting for too many points: "Make it clear what is being docked and why. Do not let the 'English professor' style of grading exceed 10% of the memos grade because the purpose is primarily to convey technical learning, not to co-enroll in writing 200. Therefore, the 'do you know how to write' component of the grade should not outweigh the 'have you learned what you have learned in engineering' component." Interestingly, this student speaks with authority on the purpose of the memo, which was to "convey technical information." In fact, the purpose of the memo was not to convey the technical, but to learn the conventions of the genre. In this particular case, the writing and its context were actually *more* important. Inherent in this comment is a level of ethnocentrism such that an undergraduate engineering student felt compelled to speak authoritatively on the purpose of a class assignment, when, in fact, students have little knowledge of assignment development (e.g., Godfrey and Parker 2010). What they *do* have knowledge of is the learning objectives associated with the course. Despite clear communication about specific writing objectives, it seems students make sense of course assignments and other classroom activities according to the cultural aspects of being an engineer and doing engineering, while distancing themselves from soft skills such as writing and communication. We see further evidence of this in students' commentary about the tension between communication and engineering.

Students expressed resentment toward communication because it detracted from engineering work, as these typical comments illustrate: "I'd like engineering to be engineering. If I want communication instruction, I'll take a communication class." "They need to introduce their own separate classes. All of the writing assignments really take away from the original course content." As seen in these comments, communication instruction was considered a distraction from engineering material. Despite the importance of communication to professional engineering success, students expressed a desire to learn communication on their own terms:

> "If I cared about communication, I'd take a class on it." "I took a communication and writing class to learn this stuff. I consider my knowledge here sufficient."

For students, real engineering is math, science, and design. Speaking and writing are incidental skills that only impede the real work of engineering. So, despite an understanding of the role of communication in professional engineering success, students seemed to want compartmentalized instruction in communication that they could decide to embrace at their discretion.

Engineering students' socialization and early identification as engineers impact students' views on the utility of communication and perceptions about the authenticity of instruction, with the potential to translate into real resistance. They privilege tangible, quantifiable concerns that have practical relevance and somewhat devalue communication instruction and consultants. Underlying their views on utility and authenticity, and ultimately resistance, is the hierarchical ordering of engineering over communication, stemming from a norm of masculine privilege that has been reinforced through disciplinary history and enculturation.

INSTRUCTOR THEMES OF ENGAGING WITH THE FEMININE IN THE CLASSROOM

We saw two prominent themes emerge when coding the data for students' engagement with the feminine: (1) the sexualized female and

(2) disciplining of the feminine. Comments characterizing these themes refer more directly to the feminine physical presence, rather than to the feminized content area. As Godfrey and Parker (2010) point out, engineering students are inclined to accept difference when "different" resembles a typical engineer. That is, difference in terms of gender is accepted provided the conditions of being an engineer are met (i.e., knowledge and competence). This section examines what students do with the "other," the one who is dissimilar from the majority: the female body or feminine performance. Data analysis highlighted a consistent devaluing of the feminine in various ways. These examples are illustrated through referencing students' unsolicited course evaluation comments that either sexualize their feminine instructor or discipline the fact that she is in the (masculine) classroom.

Sexualized Female

Traditional feminist scholars contend that commonly the female body is sexualized in order to objectify her and keep her in a box. The context of the engineering classroom, however, presents a new complex point of examination of the "other" through the female body. Primarily, within this setting, the female body is an outsider *and* a graduate student. With respect to the hierarchy of disciplines and knowledge, these bodies represent primal landscape for male objectification. From a masculine perspective, the female body is understood to be a threat to the norms characterizing the engineering classroom and to understandings about who is welcome in that gendered, disciplinary space. Thus, as the responses suggest, sexualizing becomes a way of identifying her body. The following comments are representative: "She's hot," "She's really pretty; call me," and "My friend likes you, check yes or no if you like him."

The sexualized comments (e.g., "hot" and "pretty") represent an attempt to understand or to make sense of the body in front of the room. In contrast, the "check yes or no" inquiry infantilizes (i.e., reduces power) the setting by reverting to elementary days of note passing. Consequently, sexualizing the female body as a conquerable object renders her powerless and no longer intimidating. As Atkinson (2008) illustrated, male students sexualize female teachers in an attempt to reduce legitimate power in the classroom.

Additionally, the female body is not viewed as a knowledgeable, professional body in a professional space; rather, the female form is negotiated as out of the ordinary. The emphasis students' comments placed on stereotypical female attributes, suggests that women are not to be taken seriously within the classroom. For example, when asked what else the student would like to learn about communication strategies, one student replied, "I would like to learn more pick-up lines." Here, the student comment suggests that the communication instructor can only share "soft" dating knowledge rather than knowledge of substance in the "hard" setting. In another example there is reification of the lack of professionalism in response to the female body in the situation as illustrated by this student's comment: "I would like to learn how to communicate orally in bed." This student praises the sexual connection to the female body, equating soft knowledge with female body, and something to be playfully disregarded. Similarly, in response to a question asking what students learned from the writing instruction in the class, a student questioned, "Did you have sex with my TA?" We see that even at the forefront of their education, early enculturation appears accepting of students' desire and tendency to disregard their female consultants, because they are better left objectified and pronounced as "other" in this space. Taylor (2013) argued that "sexualization devalues the female body as a way of making sense of her" otherness (8). Therefore, these responses to the female body keep her in a subordinate position insofar as hierarchy is maintained in relation to knowledge, or perceived lack thereof.

Similarly, data illustrated that the female body is one more object to be trivialized and conquered, or in this case devalued. More specifically, the female body was regarded in an animalistic connection. To this end, attempting to compliment the instructor, one student wrote, "I like her. Keep her and giver her a longer leash." This comment exemplifies the intrinsic tension between negotiating appreciation of the female instructor and otherness. As a result, this exemplifies the very fine line between sexualizing and disciplining the feminine body through the visual ramifications of a "leash." Perhaps arguably, the students' comments from this particular class/data set suggest a fondness of the communication instructor. However, regardless of "like" or "dislike," the presence of the "other" must be negotiated through discourse. In other words, "these students refer to and trivialize the presence of the other" because the "other is powerless through [her] difference [and] she is not who[m] the students are used to seeing in the classroom" (Taylor 2013, 8). Through their discourse, students (perhaps unconsciously) attempt to maintain hierarchy by positioning the female body as subservient via views of it as a sexual and nonknowing being, thus not privileged within the classroom.

This points to a perceived threat to traditional assumptions of knowledge and the body that contain/deliver knowledge. Similar to the aforementioned material, the body is sexualized; however, if she was not "sexy," then comments about her unattractiveness were at the forefront. That is, her body does not conform to either traditional knowledge or feminine assumptions within this context. Once again the masculine bodies, through their discourse, act as gatekeepers for determining inside and outside members. The engineering culture supports the "boys club" mentality, where masculine behavior, bodies, standards, desires, communication, and knowledge are advantaged.

Disciplining of the Feminine

In addition to the sexualized female, students also provided commentary designed to discipline the feminine. Disciplining the feminine refers to instances in which disciplinary institutions keep the feminine performing within the confines of standards and roles acceptable to their masculine ideological structures (Blair, Brown, and Baxter 1994). Essentially, females cannot transgress masculinist discourses. In other words, ideological discourses promote a masculine paradigm, which subsequently disciplines the feminine through recognition of being "second to" in organizing. Furthermore, engineering students and faculty discipline the feminine, through discursively constructing the female body as solely a stereotypical sexualized being, or as less than within higher education.

Understanding the disciplining of the feminine becomes more explicit when analyzing how masculinity is treated within the engineering classroom. Culturally, engineering professors and instructors and CID/WID professors and consultants are addressed in disparate ways that maintain organizational power through labels and discourse. For example, nonengineering faculty members, such as communication instructors, are addressed casually through the use of first names only. Conversely, engineering instructors are addressed with more formal salutations, including the appropriate honorific. Gender dynamics become even more nuanced when comparing student communication around male and female CID/WID consultants. For instance, female communication consultants were made subservient to male consultant instructors as this comment illustrates: "I want to know where I can improve, Anthony is much better than the other two chicks." Even though Anthony is a communication consultant, the female consultants are disciplined (i.e., nameless and trivialized) through the student's desire for objectivity and measurable feedback.

Additionally, students did not hide their disdain for the other body in the room, especially when they felt the consultant was out of line. It is important to note that when a female consultant tried to negotiate rank through a more masculine communication style or preference toward power, students would play on feminine stereotypes in order to keep the female body in rank. Most traditionally associated with female emotions, a student wrote, "She was a bitch." Other students explained, "Lay off the power trip and lose some weight," and "I hated the little power trip the lady was on, she needs to not speak to me like I am 2." Through these two examples, the "bitch" emotion was further justified through the more linguistically specific term "power trip." In the end, it seemed that when students did not like their female instructors, responses focused around examples such as this: "All of her exercises felt like getting kicked in the balls." Essentially, this demonstrates through a derogatory claim that female presence and instruction, when attempting to participate in masculine discourse, is emasculating in the disciplinary culture of engineering.

Furthermore, one student explicitly noted the tension between different communication styles within female bodies, "Tiffany was hot, Rachel was mean and ugly, and [that's] probably why she isn't married." This provides an exemplar bridge for the sexualized and disciplined females, underscoring that the other that is even less accepted than a female body is a female body communicating in stereotypical masculine ways, thus, presenting an interesting paradox. Karen Tonso's (2007) research illustrates the double bind that females face with respect to their presumed intelligence, physical attractiveness, and expected communication style. With respect to communication instructors in this context at least, we see a disciplining of the feminine when normative gendered communication expectations are violated.

In sum, the themes around the instructor demonstrate that while engineering culture presumes an acceptance of difference, it rejects or downplays those who are *too* different. These instructors are assumed to be performing tropes of either sexual objects or the bitch in the room. Engineering students seem to make sense of their female consultants through stereotypical understandings rooted in their engineering identity and ways of thinking, knowing, doing, and being in the world.

IMPLICATIONS AND RECOMMENDATIONS

Our purpose with this research was to understand students' reactions to gender within communication-engineering collaboration in order to enhance the experience for both in-group and out-group members. Our findings illustrate that students' reactions to communication instruction and consultants are related to students' enculturation into an engineering community practice. Embedded within their responses, we see elements of an engineering way of thinking and doing that privileges tangible, measurable, relevant, difficult work. In short, students' identification with the masculine ideology of their discipline is pervasive in their reactions to this interdisciplinary endeavor. Where the common boundary object of communication could be uniting, we see evidence of division and hierarchical ordering. In short, what the students want (hard knowledge) is not what they get (soft knowledge), though this bifurcation is unproductive. Their response to this incongruence included resistance to instruction and feedback, sexualization of female experts, and disciplining of the feminine.

We see three key implications resulting from this work: (1) gender hierarchies, though perhaps invisible or unconscious for the students, pervade their enculturation process and foster ethnocentrism and stereotypes; (2) ethnocentrism and stereotypes appear to get in the way of establishing common ground with the students through the boundary object of communication; (3) and the deeply entrenched disciplinary values make it difficult for out-group members to be accepted by the students of the in-group culture—students have distrust for communication instruction delivered by out-group members, even when faculty insiders endorse the instruction.

Interdisciplinary collaboration between humanities and engineering has the potential to offer many opportunities for rewarding connections. Collaboration of this sort will likely continue and grow because communication competence and interdisciplinary teamwork are requisite skill sets for practicing engineers. Given that, we offer the following recommendations for communication practitioners working at the intersection of disciplines. Since much of students' uneasiness with communication appears to stem from ethnocentrism and a belief in the superiority of engineering over communication, it is important to foster a relationship built on trust (Lim 2003). We have created a culture of trust with our engineering faculty members, but truly developing a level of trust with students will require additional effort.

One way to foster trust and break down the myth of superiority is to rely on practicing engineers to add credibility to the communication instructors, especially in the early years of the curriculum. This is important from a socialization perspective in that we need to dispel some of the myths that are perpetuated as students navigate the engineering curriculum. Communication with students early and often is the key, as well as relying on disciplinary experts to both explicitly and implicitly add credence to the value and authenticity of communication instruction. For example, communication consultants should be introduced as part of the instructional team on the first day of each class.

They should begin to develop relationships with the students through providing information about themselves, their experience, educational background, and credibility. Disciplinary experts can further validate communication consultants' knowledge by explicitly referencing their credentials as appropriate. By affording communication consultants the responsibility for providing instruction and feedback on writing and speaking, engineering faculty will implicitly acknowledge their credibility. The objective should be to create a partnership among the instructional team that is presented as a collaboration capitalizing on individuals' strengths to best meet the students' educational objectives.

In addition, communication consultants should increase contact time with students beyond dedicated class sessions. They should seek out opportunities to work with students individually so students can begin to see consultants as individuals who are trustworthy experts in the field of communication with knowledge that will be useful to them in the classroom and the workplace. One-on-one interactions, though time intensive, present a powerful opportunity for students to develop an interpersonal connection, something the students value and desire, as Godfrey and Parker (2010) remind us. In addition to required writing and speaking consultations, we recommend that communication instructors present additional, optional learning opportunities for smaller groups of students. For example, they could offer workshops on topics not covered in the classroom or provide additional opportunities for more focused tutoring on speaking and writing. Although not necessarily tied to specific class activities, it is also recommended that consultants be "present" in engineering students' classes (even when not lecturing or facilitating a session) and spend time in students' classroom buildings to facilitate more informal communication and community building.

Beyond the divisiveness of ethnocentrism, we saw evidence of stereotyping in many of the students' comments, specifically, stereotyping with respect to the limited skill, abilities and benefits of communication consultants. To combat stereotypes, communication consultants should practice mindfulness (Lim 2003), examine their own assumptions, and be daring and engage students in a discussion of the assumptions they hold about communication. While this may feel risky for some, we believe that it is important to be direct; engage and confront the stereotypes and work to dispel them. At the same time, we must be open to sharing stereotypes we might carry into the classroom about engineers. This demonstrates a level of vulnerability that will only serve to foster a deeper level of trust.

Finally, it is important to recognize that stereotypes and ethnocentric views will likely dominate initial interactions/encounters between in-group and out-group members because anxieties run high. In other words, students are unsure and afraid of communication, and thus they experience dissonance and potential identity threats. As a result, ongoing mutual clarification and perspective taking are paramount.

Our analysis shows how students' responses to communication instruction and consultants are related to their immersion in engineering culture and the implicit gendering and hierarchical ordering of disciplines. These results present an interesting foundation for future work; however, we acknowledge limitations to this study and offer recommendations for further exploration. This research relied on anonymous student comments to make sense of gender, hierarchy, and students' experiences with communication instruction and instructors in the engineering classroom. As such, we acknowledge that students' views at the end of the term presented in an anonymous format may have impacted their sense making. Thus, we recommend additional exploration of students' experiences through in-depth interviews that allow for follow-up and greater explanation and probing. This work also relied on students from only two engineering departments. To better understand the broad culture of engineering, we recommend future research that explores students' experiences with communication instruction across a diversity of departments and instructional approaches. This will allow for the development of broad understandings of student experiences with communication and the impact of the nuances of engineering culture on interdisciplinary education. Finally, we acknowledge the absence of engineering faculty members' voices and recommend that future research explore their perceptions on interdisciplinary collaboration and the impact of disciplinary cultures and ideologies on WID/CID partnerships.

This work highlights the challenges characterizing writing and communication in the disciplines stemming from deeply held ideological beliefs that inform disciplinary values, practices, and gender dynamics. The gendered tensions we illustrate appear to stem from students' reactions to identity challenges as their assumptions and expectations for preferred styles of teaching and learning are upended. Given the contact of the masculine culture of engineering with the more feminine culture of communication, this is not surprising. Yet, we anticipate that our reactions to each other, given our suggestions, might be reconsidered and recuperated.

NOTE

1. The Accreditation Board for Engineering and Technology (ABET) mandates that engineering undergraduates demonstrate competency in both technical and professional skills upon graduation.

9
INTERCULTURAL COLLABORATION
Respect, Relationship, Responsibility, and Reciprocity

Sundy Watanabe

Finding creative, integrative solutions . . . in . . . problem-filled circumstances is no mean accomplishment. (Smith 1990, 274)

Scholars Katherine Schultz and Glynda Hull explain that the vocabulary commonly used to talk about literacy—that is, events, practices, and activities—has arisen from a flurry of earlier scholarship that falls within three major theoretical categories: Ethnography of Communication, Vygotskian Perspectives/Activity theory, and New Literacy Studies (NLS) (Schultz and Hull 2002).[1] This is important on three counts. First, it reaffirms that literacy is situated within social and cultural contexts and that experiences out of school influence those in school and vice versa. Second, it reiterates that actions surrounding the enactment of literacy are constructed in multiple ways, emphasizing "purpose within context and the patterned interplay of particular skills, knowledge, and technologies" (20). Because we understand these two points, NLS scholars today also understand a third point, that any schooled activity involving literacy is "motivated by larger purposes and aims than literacy itself" (21) and must necessarily include a discussion of power.

The necessity of accounting for power relationships adds important terms to our literacy vocabulary, terms such as Discourses, identities, and ideologies (Fairclough 2001; Gee 1996; Street 1995; Wortham 2006). We have come to understand that Discourses have little meaning outside of the contexts wherein they are enacted, the events or activities that make them useful and which they in turn influence. We also understand literacy as an ideological practice that cannot be constructed as a neutral skill set. As Mike Baynham and Mastin Prinsloo point out, "literacy practices are always and already embedded in particular social forms of activity," and they are "fundamentally mutually constructed and . . .

DOI: 10.7330/9781607328032.c009

shaped by both institutionalized and informal relations of power" (2001, 83–84). In the academy, literacy practices can often be used as gatekeeping devices to wield power, establishing inclusion and exclusion, in-groups and out-groups, equal participation, and less or none at all.

The singular direction encoded in what we think of as literacy education reproduces certain kinds of workers and employers, students and teachers, as well as delineated sets of values and knowledge systems. The singular direction of these practices is considered and made normal, or normative. Normative literacy practices tend to privilege and maintain Euro-Western expectations of schooled performances: teacher as expert, student as novice or apprentice, lecture as the primary means of delivering content, writing as the dominant mode of conveying knowledge, learning as an individual accomplishment demonstrated through examination, and skills taught in isolation of meaning making and community building. Somehow, authority figures of all eras generally think they possess the truly important knowledges and are therefore obliged to direct and control delivery of those knowledges, downplaying and sometimes outright dismissing "other" knowledges as lesser and/or deficient. Literacy practices can be used—even unconsciously—as a way to establish power, as a means of control. While no longer ascribed to by most literacy scholars, normative literacy practices continue to undergird much scientific learning within higher educational systems and, hence, to reinforce the status quo, reinscribing difficulty and failure for groups who wish to practice or perform differently.

The aim and consequence of such educational practice (and attitude) are not just socialization but assimilation and regulation, keeping power (and the powerful) in its/their "rightful" place. When assimilation and regulation are not thought possible or desirable (or desired in some cases), othering and ostracism can occur. In such cases healthy learning relationships are impossible to establish or are relinquished. While fully acknowledging the challenge presented, in this chapter I affirm that successful learning relationships and productive intercultural collaborations can happen, even between those holding disparate cultural knowledges. Specifically, I argue that engineering and humanities collaborations can benefit when sojourners—those residing and having extended experiences in another's disciplinary setting—accommodate broader notions of power and knowledge. However, the benefit I wish to talk about only becomes possible when collaborative partners willingly employ shared power and knowledge, when they accommodate broader notions of what constitutes power and knowledge in the first place, and when they imagine their interactions differently.

Linda Flower (2003) notes that the perspectives we operate under are "inseparable from their distinct modes of representation" (62). In other words, our attitudes or dispositions cannot be separated from our ways of doing literacy, the tools we choose to "do" or represent literacy with, and the products that result from our doing and representing. If we accept that perspectives are inseparable from practices, then we also understand the need to seek "radical alternatives" (50) and to use "difference intentionally" (42). To this end, I have found it useful to draw not only from theoretical foundations found in New Literacy and Composition Studies (Archer 2006; Berry, Hawisher, and Selfe 2012; Faigley 1999; Haas 2010; O'Brien and Eriksson 2010) but also the theories and methodologies found in Critical Indigenous Studies (Brayboy 2005; Brayboy and Castagno 2008; Brayboy and Maughan 2009; Brayboy et al. 2012; Kana'iaupuni 2004; Riecken et al. 2006; White et al. 2002). I bring these knowledges together not to conflate them but because through them we can think about engaging collaboration differently, perhaps radically so.

A Critical Indigenous Studies[2] approach can infuse collaborations with a structure that brings to the fore the ways knowledge is acquired, exchanged, and valued. Such an approach can make visible the discriminations that work to privilege Euro-Western literacy practices. It can also encourage collaborators to disrupt their accustomed ways of doing in order to create "alternative images" and "contradictory visions of outcomes" (Flower 2003, 56). One shake-up of the familiar leading to alternative images, for example, might be embracing a holistic rather than a discrete or silo perspective. Many Critical Indigenous Studies scholars, for example, make no definitive separation between ways of being (cosmologies), valuing (axiologies), and knowing (epistemologies). Manulani Meyer (2001), in fact, states, "There is no such thing as isolation from the rest of creation, and . . . this relatedness provides a basic context within which education . . . occur[s]" (145). Since all life experiences produce and inform knowledge making, Critical Indigenous Studies scholars often use "epistemology" as a comprehensive term, and they do so to counter deficit views and confront what Michael Marker (2004) calls "historical imbalances" (20) in educational contexts.

Ray Barnhardt and Angayuqaq (Oscar) Kawagley's epistemological work identifies specific values, which they call the Four Rs, namely, respect, relationship, responsibility, and reciprocity (2005). Other scholars also include relevance and reverence (Ball 2010; 2011; Cajete 2005; Kimmerer 2002), but for the purposes of this chapter, I follow Barnhardt and Kawagley in focusing on the first four. These values are not only directly useful for collaborations but are also fundamentally practical in that they work for the benefit of the larger communities wherein they are enacted. The Four Rs help collaborators and their respective communities think carefully about how their work is influenced by shared and separate histories, how in this sense it is—and they are—related. Relationality amongst collaborators and communities means understanding each is "part of a larger cosmos, not the center of it," and each is responsible for respectful interactions because these will have "long-lasting repercussions" (Brayboy et al. 2012, 438). A Critical Indigenous Studies approach, then, necessitates a collaboration that will "care for both the ideas, or knowledge, it generates and the living beings [its] ideas influence" (438). Such an approach obligates reciprocity as manifested in interactions, where collaborators listen to one another with more depth and humility and, thereby, honor relationality (see "all our relations" in Wilson 2001, 177).

A CRITICAL INDIGENOUS STUDIES APPROACH

In this section, I contextualize what I argue above: incorporating a Critical Indigenous Studies approach has potential to increase powerful collaborative relationships. First, I summarize select points made by Indigenous scholars Nancy Allen and Frank Crawley, Gregory Cajete, and Steven Semken, who work in scientific fields. Their points, as summarized here, illustrate epistemologies distinctly different from Euro-Western norms. They tell us, by contrast, that dominant understandings of science are, in fact, culturally constructed and culturally specific. Second, I retell Barnhardt's (2002) story of instructor and student interaction in Alaska to exemplify the way Indigenous epistemologies are interwoven in a real-life collaborative experience. Taken together, the work of these five scholars leads me to suggest ways that normative literacy practices associated with collaboration between very distinct disciplines can be reimagined and shifted toward an alternative pedagogy through the Four Rs mentioned previously: respect, relationship, responsibility, and reciprocity. As Barbara Monroe (2015) attests, "Language use is not just linguistic and discursive but also epistemic and semiotic, encoding a people's epistemology. As such, it carries forward not just a rhetoric but also a worldview" (38).

Allen and Crawley's worldview theory holds that existing knowledge structures the way we receive, process, and formulate new information and gives shape and coherence to the way we relate to our environments (1998). Within higher education, they observe, scientific disciplines presuppose shared epistemologies and ontologies: Euro-Western ones. They are troubled by this presumption because they find distinctive differences between Indigenous and Euro-Western worldviews in terms of epistemology, pedagogy, and perspective. In particular, they note an Indigenous preference to learn content through long-term observation in the field "rather than recitation" of facts from textbooks (125) and written examinations. They also posit a reluctance to draw definitive conclusions early on in instructional and research processes, especially when those conclusions are drawn from short-stay experimentation within a variety of locales and environments rather than long-term depth and breadth experiences within one context. The differences between communities' knowledge-making work can seem stark, but Allen and Crawley provide the caveat that while worldview theory provides a powerful way of understanding difference, it is important to remember that the differences noted are not truly "inherent in either science education or participation in the science community" (129; see also Cajete 1999; Lipka 2002; Swisher and Deyhle 1989). Still, they advise, science and technology would do well to become less fixed in their Euro-Western paradigms and more open to other worldviews.

Cajete's (1999) work, in acknowledging the way American Indians utilize modes of learning to produce knowledge, argues that Indigenous knowledge making is "mediated by [a] particular cultural orientation" (165). In saying so, Cajete parallels Allen and Crawley's conclusions. However, where they suggest a "worldview" paradigm to understand difference in orientation toward knowledge making, Cajete offers another: a bicultural paradigm. A bicultural (or multi- or inter- or transcultural, for that matter) approach, he suggests, requires an understanding of Indigenous core values and how these manifest as observable behaviors. Cajete first references *respect* as a primary core value, and *relationship* becomes his last reference. Respect and relationship effectively bookend a spectrum of values, reaffirming pedagogical preference for cultural behaviors such as quiet/silence, patience, practical and purposeful work, group security via consensus and cooperation, listening and observing, relative time, holistic orientation, spirituality, caution, and regulating discipline through group pressure (150–53). When these values and behaviors are identified and applied in the context of science education, Cajete suggests they provide the energy needed for transformational learning (149) toward shared power and knowledge.

Steven Semken (2005), too, takes up the idea of cultural difference influencing knowledge making, noting that few Native students study earth and environmental science because of discontinuities caused by differing philosophies, values, and meanings that make up Euro-Western university experiences. Similar to Allen and Crawley's stated preference for long-term observations in one locale, Semken advises attending to an Indigenous sense of place, which he identifies as locally constructed and historically significant. Attention to place, he says, is purposeful, in that it overcomes cultural discontinuity by helping people attend deeply to their environments and their work with specific communities. He describes place-based education as embodied or lived, and it thereby becomes relational, spiritual, and reciprocal. If curricula and action are to be purposeful and community oriented, he advises, they should be guided by an explicit focus on the "diverse meanings that place holds for the instructor, the students, and the community" and teaching that "enriches the sense of place of students and instructor" (153). Place-based approaches, then, accommodate Indigenous epistemologies that emphasize contextualized, historical understandings and explorations.

Barnhardt's (2002) ethnographic account of an educational development program utilizing field- or place-based training for Native students and university faculty provides one specific illustration of a reciprocal intercultural educational experience contextualized by worldview, place, and bi- or intercultural relationship. Barnhardt describes the experience of an academician and a Native student learning together in the tundra of western Alaska. The academician's task was to teach the terrain of university-level reading and writing. The student's was to teach his teacher to read the terrain of the arctic environment. On one excursion, the two—both experts in their respective fields and sojourners in the other's—experienced a difficulty that tried their collaborative venture. The academician thought he could read snow conditions well enough to break his own trail, causing him to ignore yellow patches, which, the student later pointed out, should have alerted him to danger. Instead he struck a new trail, and he and his snowmobile promptly sank waist deep into freezing water. The student pulled the teacher and vehicle out of the ice and then quickly began digging in the snow for grass. The teacher, based on prior experiences and what he thought was necessary and appropriate, assumed the student would make a fire with the grass for warmth. Instead, the student instructed him to remove all his clothing and then get back into the shell of his snowmobiling suit and replace his boots. Not understanding, but trusting enough to act anyway, the teacher did as instructed, and the student quickly stuffed both the suit and the boots with the tundra grass he knew (how) to gather and productively use for insulation.

This story represents a decisive juncture in the collaborative experience of the sojourners, one that required respect, trust, and action. Luckily, in the instance described, the instructor-as-student understood the student-as-instructor possessed necessary and valuable knowledge, knowledge that prevented him from freezing that day. Reciprocally, through sharing his academic expertise, the instructor was also increasing the student's potential to survive in the world. Academic survival, while not often thought of as a life-and-death situation, can have equally consequential physical and material outcomes in terms of employment and financial security. While the student and teacher in Barnhardt's example demonstrated literate responses according to knowledge they already possessed, the modes of learning they utilized underscore the importance of knowledge displayed according to relevant community context and practical need. Community experience determined the authority to assess and implement needed action according to history or tradition, and the community setting established the criterion that evaluated or legitimated the outcome. Knowledge was "a process tied to creation" (Battiste 2002, 13–14), a "resourceful capacity of being" and not a "commodity" that could be "possessed or controlled" (15). It was "a living process to be absorbed and understood" (15).

Given what can be learned from the scholars above, I am troubled that gaining scientific knowledge, from a Euro-Western scientific perspective, still assumes normative literacy practices. Thinking of knowledge making in strictly this way reaffirms why we might experience cultural discontinuity when the way we relate to particular environments conflicts with another's relational viewpoint. However, cultural discontinuity may be overcome by attending to Indigenous worldviews, epistemologies, and senses of place (see Feld and Basso 1996 for similar arguments). In Barnhardt's ethnography, knowledge was an exchange, a point of reciprocity allowing instructor and student (acting in either/both of those roles) to experience the nuanced adjustments required while sojourning in the others' particular context. Both had to exhibit openness to difference and vulnerability. When they did so, their actions become more responsible, more relational, and less abstracted from everyday experience. Such "openness," according to Kenneth Liberman (1999), "may force one to reset or abandon one's priorities. The contingencies of field inquiry are not to be viewed as obstacles ... but as opportunities to learn which inquiries are the ones that really matter" (49).

Incorporating a Critical Indigenous Studies approach does not deny the possibilities afforded by the traditionally understood educational epistemology of Euro-Western institutions; rather, it seeks to "place" its influence within a relational context where Indigenous conceptions of knowledge making are an integral part of coming to know. Critical Indigenous Studies, as a foundation, can be utilized to change (transform) the conflicts that often arise in contact zones—when different cultural groups holding asymmetrical relations of power meet and clash (Pratt 1991)—into more beneficial exchanges. Indigenous epistemologies, when applied to engineering and humanities collaborations, can become catalysts for learning that accrues more power for both communities. In working through the sections that follow, I reiterate that exclusionary epistemological choices and perspectives within disciplines (the silo effect) can make collaborations with others difficult. But I also offer some recommendations—some alternative images, if you will—based on the scholarship noted, my own experience, and the experiences of others. Incorporating these recommendations into instances of sojourning can assist collaborators in taking steps toward real, creative, practical transformations that benefit both communities.

A CASE STUDY OF INTERCULTURAL DISSONANCE

In the contact zone of the engineering and humanities collaboration, my charge, with input from faculty members and students, was to develop and implement a four-year plan, redesign course syllabi to include writing, and present writing colloquia to faculty. In addition, I was encouraged to attend content classes when my schedule permitted, deliver classroom lectures upon request, comment on student writing assignments, and provide face-to-face consultation. This work was expected to result in research, both on an individual level of study and on a Center-wide level. Collaborating departments also conducted their own research regarding the collaboration, and the collaboration was expected to facilitate positive change. It did so more or less successfully because of distinctly different epistemological approaches to disciplinary knowledge making.

In theory and method, the engineering department favored a quantitative approach to learning and pedagogy: objective, outcome and grade based, linear, efficient, and form guided. The approach posited authority as residing in knowledgeable instructors who delivered quantifiable scientific content to students, both of whom were and are today in high percentages male.[3] Instructional delivery typically took the form of lectures. Even in what were couched as discovery situations, such as labs, students followed detailed guidelines of methods and procedures to achieve the aims of experiments. If they followed the instructions, it was understood their experiments would produce predetermined, logical, and numerical results. Countable results could occur in these settings because the objects of experiments were generally believed to behave in repeatable fashions. If they did not, it evidenced an aberration either in performance or equipment that was resolved by applying another set of logical methods and procedures. Reliance upon objectivity made sense in this context, for if "things really do exist in some prefigured constellation and operate in accordance with natural laws" and if knowledge is certain, it is "entirely appropriate to demand an objective stance"; otherwise, results are "distorted" and take on "unreal characterization[s]" (Guba 1990, 87). However, if a discipline accepts both the "physical and the nonphysical realms as reality" as Lynn Lavallée (2009) tells us many Indigenous communities do, "one must accept that reality cannot always be quantified" (23).

The disciplinary history guiding the humanities arm of the endeavor deliberately worked from a qualitative approach. Those of us working in the Center understood inquiry to be socially constructed: more intersubjective, rhetorical, and process oriented than objectively measurable. It was recursive rather than linear, more discursive than form guided, and more messy and unclear than efficient. This work was (and often still is, for good or ill) characterized as feminine and of lesser value, as Kedrowicz and Taylor illustrate in the previous chapter. While authority was still mostly hierarchical, it appeared less so than in engineering. Humanities collaborators assumed interactions wherein individuals or groups would work from difference toward accommodating various outcomes, whether consensus/agreement or dissensus/disagreement. In practice, objectives were difficult to pin down and hardly ever predictable, especially in regards to time lines. Since the objects of study were humans and human behaviors in writing, applying any set of methods and procedures hardly ever resulted in predetermined, let alone repeatable, results. Indeed, if that happened, the research was considered suspect; perhaps, the thinking went, data had been manipulated.

Understanding these very different histories and approaches, it was no wonder the faculty and students with whom I worked put up some resistance, preferring to study electrical cells or currents rather than rhetoric and writing. Sherry Turkle and Seymour Papert find that intellectual approaches to "hard" science are based on abstract thinking and systematic planning, that is, dissecting things into parts and then systematically planning how the parts work to make the whole (1990). Leah Buechley and Hannah Perner-Wilson posit these epistemologies are then "built into the standard *tools and techniques* of electronics" (2012,

17 emphasis added). Conversely, Turkle and Papert find, "soft" science approaches favor iterations of negotiation and renegotiation based on concrete forms of reasoning, for example, talking about alternatives and moving things around until a satisfactory design is achieved (1990). These, too, are built into preferred ways of making, like revising an essay or a report by moving sentences and paragraphs around or like making a poster by moving words and images around until the product is aesthetically pleasing and visually comprehensible.

Since expertise within the "hard sciences" has long assumed powerful precedence, no wonder I, a rhetorically influenced writer and qualitative researcher, was a bit rattled by the specter of strict adherence to empirical method, including prioritizing objectivity, validity, and generalizability. On a conscious level, I understood that human beings derive methods and procedures (Eisner and Peshkin 1990); methods and procedures are neither good nor bad neither effective nor ineffective in and of themselves. Audience, purpose, and angle of vision determine which methods are used and how procedures are followed. I knew tables, graphs, and mathematical equations constituted one valid way of viewing and knowing and proceeding in the world, one way of recording and categorizing knowledge. However, it seemed equally valid to assert the benefit of paying attention to rhetorical situations and incorporating invention activities and interactional discussion to teach writing because methods "based upon a deterministic causal model simply do not fit the arenas in which human interaction takes place" (1990, 11). In other words, such methods were not applicable in the arena of human interaction and communication.

The point was that based on a host of scholarly studies, I knew the binary constructions of knowledge making established in these disciplinary houses were extremely problematic. Singular opposing perspectives limit how one perceives human beings and their activities, in this case writers/writing within an engineers/engineering setting. I also knew that those who write or study writing would hardly say the process was entirely linear, efficient, or form guided even though the end product might allow it to look so. And, truly, in practice the same could be said for engineering processes. Even though the methodical scientific approach to a lab procedure might allow the end product to look neat and tidy, the product belies the process. Theoretically, methodologically, and experientially speaking, I understood this, so speaking persuasively about differences and similarities should have been easier. Sojourning should have been easier. Still, I had a hard time translating my knowledge into the kinds of communicative products accepted by the engineers, the kind of data they called valid. They, too, had a hard time accommodating a different worldview, a hard time accommodating change, even that which would be ultimately beneficial for both communities.

The engineering instructors with whom I collaborated knew the value of scientific lab time. Yet, they undervalued the ways similar, devoted time might improve writing skill and at the same time facilitate content knowledge. Engineering students were smart, they joked. Smarter than humanities students, they implied. A "how-to" handout was all they really needed, and maybe an assessment checklist. Hence, the faculty's priority was to fill a website with one-page handouts of writing instruction—how to write an introduction, how to write a conclusion, how to participate in peer review, how to write a memo, how to describe a process, how to write a lab report, and so on—and, having done so, call the engineering and humanities collaboration a success. My enthusiasm for this priority was, as you can imagine, hardly overflowing. Instead, I felt as resistant on my end as they did on theirs, especially when engineering faculty insisted on collecting numerical statistics, or hard data, to prove to colleagues and students that the writing assignments and tasks were, as they often said, "successful and repeatable." I resisted the idea that explaining a writing task to students in the form of a lecture and handout would result in the same level of success and knowledge that hands-on practice and discussion would produce, and less efficiently at that. I appreciated that if students were only required to "fill in the blank" template forms of "how-to" documents, then, perhaps, their knowledge might indeed be repeatable. I questioned what would happen, however, when they were required to construct their own forms, using their own critical thinking, based on principles they had neither explored nor practiced.

Feeling the insecurities of hierarchical status arrangements, a graduate consultant working with faculty in their territory and upon their terms, I capitulated. I delivered the requisite pages of documents, taught forms in order to contain student writing, as the faculty had requested. But all the while, it felt slightly unethical. It constituted a dilemma between the integrity of my disciplinary perspective regarding the way writing "should" be taught and the way it could be taught according to the value system of the engineering department. It was difficult to believe the situation was in the best interest of the students. In fact, in the long run, I felt it might even be dangerous for the engineering department as well as the engineering and humanities collaboration to cede to a quantitative ontology without regard for other ways of knowing and valuing. The very results the department hoped to account for might never materialize. Faculty could then blame writing consultants or the humanities collaboration and never see that the inability to accommodate differing systems may have caused all of us to step on our own toes, so to speak. After attempting to initiate a collaborative effort between humanities and electrical and computer engineering, I was beginning to understand the difficulties and limitations inherent in trying to infuse the modes, content, and assumptions of one disciplinary domain into another.

I am not sure how clearly, if at all, the engineering faculty saw the dilemma. Certainly, they did not see it as I did, particularly since we didn't agree on what constituted valid evidence, goals, and practice. On the other hand, we did achieve some measure of success. By the end of the year, we had designed a four-year plan incorporating writing into the curriculum design, and we had begun to implement parts of it in parts of classes. When I left my position within the collaboration to take on a new responsibility and opportunity, however, I felt somewhat disconcerted. As Allen and Crawley note, "Values, like rules, are more obvious when a conflict occurs" (1998, 128). The tense undercurrents running through the sojourning experience, subtle though they were, left me thinking about contradictory visions of outcomes. What could I/we have done better? How might the interaction have been instigated, conducted, and supported so that it was more productive, more creative, and more rewarding?

STORYING TOWARD MIDDLE GROUND

In trying to better understand, I turned to story. According to Bryan Brayboy's (2005) definition of Tribal Critical Theory, theorizing through story can do purposeful work for community (see also King 2003; King, Gubele and Rain Anderson 2015; Powell 2012; Watanabe 2014). Through storying, we can enact "sense-making" as "a joint rather than a single-handed adventure" (Scollon and Scollon, qtd. in Monroe 2015, 23). Storying, therefore, works collaboratively toward shared power and knowledge. Consider the story Janine Benyus (2005) relates about her own collaborative experiences.

Benyus, a science writer and innovation consultant, saw a disconnect between biologists and the architects, designers, and engineers who, in her words, "make our world." The two groups, she said, were "not talking to each other. At all." Thus, they were missing opportunities to solve some difficult problems. One of those problems was scaling, the buildup of calcium carbonate material inside pipes that eventually closes them off. Wastewater treatment engineers had two options to deal with scaling, but neither was very good, and progress toward other solutions had halted. Benyus and a group of biologists invited the engineers to come to the Galápagos and look at what "biomimicry" had to offer them, but the engineers were resistant. "We already do biomimicry," they said. "We use bacteria to clean our water." The biologists reminded them that, no, "bioprocessing" was really just bioassisted technology. Biomimicry, on the other hand, involved learning directly from organisms and applying what was learned to design technology. The engineers decided to go.

When they got to the islands, the engineers still didn't realize what could be learned. They looked around; they took some pictures. Seeing the engineer's lack of active engagement, Benyus picked up some shells to show the engineers. She pointed out that the shells were made of calcium carbonate, the same material as the build-up inside the pipes the engineers were concerned about. Seashells, she said, are "templated by proteins, and then ions from the seawater crystallize in place to create a shell." The same process without the proteins, she suggested, was happening inside their wastewater pipes. Okay, the engineers said, that's interesting. But why doesn't the scaling on the shells continue to enlarge, as it does in the pipes? Benyus answered: "The same way seashells exude a protein and it starts the crystallization . . . they let go of a protein that stops crystallization." That protein adheres to the face of the crystal and stops its growth. In fact, she said, scientists had developed a product called TPA to mimic that "stop-protein," creating an "environmentally friendly way to stop scaling in pipes." And that, Benyus says, is when everything changed.

Suddenly, the engineers "got it," and an instance where sojourning initially produced tension transformed into an instance of productive knowledge making. The engineers saw the value of collaboration, and they employed alternative images that worked toward contradictory visions of outcomes. They quickly shifted from passively looking at the environment to actively inspecting and acutely observing its natural organisms and processes. According to Benyus, "Learning *about* the natural world is one thing. Learning *from* the natural world—that's the switch. That's the profound switch. What [the engineers] realized was that the answers to their questions are everywhere; *they just needed to change the lenses with which they saw the world*" (emphasis added). Benyus's lens was collaboration that works toward environmental activism, helping design and technology engineers see the world through the eyes of a biologist in order to innovate in a more nature-friendly way. As Benyus notes, for the engineers and biologists the context was the earth, "the same context that [they were] trying to solve [their] problems in." Like the instructor in Barnhardt's Alaskan context, the sojourning engineers came to understand what—in some contexts, in some places—they did not know, and they came to know differently.

We, too, can understand and know differently because of Benyus's collaboration story. We can interpret it from a Critical Indigenous Studies perspective, which tells us that active, receptive listening and observation allowed the collaborative partners to answer an inquiry that mattered. Because the collaborators enacted knowledge from a new perspective—we could say reciprocally—the solution to a weighty problem appeared. To get to that point, however, the engineers had to lower their defenses about what they thought they already knew. They had to trust something might be gained by collaborating and sojourning within another's place and context. Both collaborating parties, engineers and biologists, had to respect another's way of being, doing, and knowing. Although both parties were expert in their own spheres, they were novices in the other's; each had something to learn and each had something to teach. Then, too (as in Barnhardt's story), their perspectives changed as they worked on the problem in a localized context, specific people holding particular histories coming together in a particular place. The context allowed them to understand differently and thereby accrue more problem-solving power. Bringing together their shared power and knowledge, they created something better. They came to better understand the ongoing relationship between (inanimate) organisms and humans.

Robin Kimmerer (2002) asserts, "A thing is understood only when it is understood with all aspects of human experience, that is, the mind, body, emotion, and spirit" (435). Understanding is a difficult thing. I am certain I have not fully accomplished it. But now, a few years out from my sojourning experience in engineering, I know I better understand the importance of critical reflection: bringing mind, body, emotion, and spirit to bear on worldviews or perspectives. I better understand how difficult it is to strive for harmonious, respectful practices when, inevitably, tensions arise; yet, it is an ethical responsibility. And as I address *this* ethical responsibility, let me be clear about another responsibility concerning the theoretical and methodological lens I bring to bear on the problem. In applying a Critical Indigenous Studies approach to my sojourning experience, I do not wish readers to make facile and false correlations between the relative marginalization of humanities / relative privileging of engineering and the very real consequences colonization has exacted upon Indigenous nations—historic and contemporary, material and spiritual. It is precisely because educational systems, colleges, and departments participate in (re)producing colonialisms—normative literacy practices, in this case—that it becomes necessary to question the "historical forces shaping societal patterns as well as the fundamental issues and dilemmas of policy, power, and dominance in institutions, including their role in reproducing and reinforcing inequities" (Marshall and Rossman 2006, 6). A Critical Indigenous Studies approach helps us do this necessary questioning and moves us toward more productive and respectful relationships.

Attempting to enact collaboration places sojourners within liminal space—between, or on a middle ground[4]—and is a positioning some might disdain as playing it safe. They see it as not taking a position. I see it differently. Consider the way the words "middle" and "ground" can be defined, and the force of this placement becomes apparent. Middle can be defined as a point of convergence, the point at which everything is connected and bound. It is comfortable, but it can also feel confining. It can connote wavering and indecision but also solidity and firmness. Meeting in one central place could be considered safe, yes, but it also implies the risk and benefit of making choices. Hence, middle can be viewed as a point of departure. Every choice, every moment of shift at every disciplinary border or boundary crossing, becomes a moment of possible disequilibrium and disagreement or a moment of creative and productive change.

Middle (between) is a juncture that allows (requires, in fact) flexibility. It is a tethering spot for both flight and return. The overtones and juxtapositions of opposition and possibility could continue indefinitely. Ground is of the earth, a basic element. Life depends on it. Indeed, disciplinary "houses" sit on it. Its synonyms include "foundation," "base," and "root." Add an "s," and the definition shifts toward evidence, cause, and purpose. Add an "ed" and, in electrical terms, it means to harness a current, to focus it at one point to make that which is highly volatile stable and able to be contained. A grounded current, while potentially dangerous, can be plugged into for electricity and energy. If one thinks, then, of sojourning as coming to a middle ground, one plugs into or is positioned at the point of intersection where there is a balance of negative and positive energy. A key in both electricity and Diné epistemology is the balancing of these contradictory forces, as it keeps the universe in motion, in harmony (see Toth, personal interview, February 12, 2015; see also Brayboy et al. 2011). Collaborative sojourners, we might say then, engage at that point of electricity and energy, at the conductive point surrounding shared interest (see Deloria 1970 on "tangent lines," 12; and Watanabe 2012) to produce a transformative experience.

Brian Burkhart (2004) teaches that real connectedness is imperative to intercultural collaboration. Merely "forming an alliance or making a treaty," as Vine Deloria (2001) cautions, "does not address irreconcilable differences in worldviews" (8). We must, Burkhart (2004) says, "maintain our relations, and never abandon them *in search of* understanding, but rather find understanding *through* them" (25, emphasis added). He explains that Euro-Western philosophic conceptions began with Plato and are primarily concerned with figuring out what man is and how he is differentiated from everything else. Euro-Western theory or philosophy is mainly a thinking process, evocative of a desire for centralized and "permanent" answers to hypothetical questions. It asks that truth be made evident via reason and empirical evidence. Conversely, Burkhart says, American Indian philosophy is more concerned with questions that evidence the right ways for humans to act (not just think), and it considers how those actions are related to rather than separate from everything else in the world, whether animate or inanimate. Knowledge comes through lived experience that seeks "a way of seeing the whole" (25). Indeed, relationship as/to the whole—*we are; therefore, I am*—cannot be divided into categories or differentiated into parts—*I think; therefore, I am*—as per Descartes. Cajete observes, "The people are an ear of corn," not individual kernels (cited in Burkhart 2004, 26).

Euro-Western approaches to education accept that for the purpose of answering speculative questions about truth, knowledge is differentiated into categorical branches: that is, literary, philosophical, and scientific. But, truly, this system makes little sense from an Indigenous worldview. Relational knowledge resists the force of procedural and distantiated method. Instead, knowledge is accessed along many levels and is acquired as needed, in just the right amounts for the task at hand. It occurs when the knowledge seeker and knowledge conditions are adequately prepared and ready. Because "all investigation is moral investigation," this readiness factor places limits on the timing of questioning and is guided by the radical idea that "certain things should not be known" (Burkhart 2004, 17) just because someone is curious at the moment. Demanding knowledge or "proof" is not always right (20). Humility, openness, and patience, not merely inquisitiveness, are required when coming to know. Knowledge comes with an attendant responsibility, and so perhaps "more knowledge is not always better" (18). Knowledge comes when one can responsibly act upon it. This epistemological sense is important to knowledge making because "how we behave ... in a certain sense shapes meaning, gives shape to the world. In this way, what we do, how we act, is as important as any truth and any fact" (16–17). Knowledge, then, is not gathered as much by distancing oneself from the surrounding world in order to know as it is by actively observing and participating in what is directly at one's feet—whether that is tundra grass, a seashell, a memo, or a circuit board—in order to be, know, and do (see also Kirkness and Barnhardt 2001; Barnhardt and Kawagley 2005; Lyons 2010).

CONCLUSION: REIMAGINING COLLABORATION

If we sojourn, we cannot stand apart and pretend another worldview does not exist. The history engineering and humanities share, by virtue of academic context alone, makes connection inevitable. We might, however, more strategically and deliberatively choose to enact greater connection using Critical Indigenous Studies approaches. Powell (2002) reminds us that "scholarship is an act of imagination" and that when confronting notions of normative literacy practices within Euro-Western structures we can work "imaginative liberation" (399). This may mean strategically acquiescing to Euro-Western epistemologies, but it certainly means critically questioning them from the position of strength. Imagine the making of a lab report, for example.

Rather than mandating a lab report form that insists on one right way—individual, systematic thinking, and dissecting the document into parts—an engineering and humanities collaboration might enact strategic and deliberate connection by encouraging community dialogue toward multiple approaches, including arranging and rearranging possible parts to make a whole. Imagine this process beginning well before a class meets, or before a term starts, for that matter. Envision faculty members and writing instructors meeting with community elders, brainstorming together from the perspective of their respective expertise and recognizing in the other shared knowledge and power. Imagine that as they research what is known and what needs to be known, they prepare questions that need to be answered and explore problems that need to be solved. Then, picture an engineering faculty member and a writing instructor standing side by side introducing those questions and problems to a classroom of students. Can you see them facilitating whole class and small group discussions concerning history, definitions, expectations, and the audience and purpose their lab reports will serve? Can you imagine them standing beside students at whiteboards or sitting with them at computers as they question, research, summarize, draw and model, observe processes in the field, and finally propose possible templates? Later, imagine them experimenting with various templates during their own lab work, and afterward storying together, discovering and explaining problems they encountered, so they can reformat the template/form to better serve its intended purpose. Imagine that at every turn collaborators exhibit respect, responsibility, reciprocity, and relationship/relationality.

Yes, such a process takes good amount of time and a good amount of listening and processing. It should. It must, if it is to take into account not only current moments, teachers, and students but also those coming in the future. It takes patience for collaborators to negotiate consensus or acknowledge dissensus. It takes trust to bypass the supposed efficiency of lecture, trusting that content will still be taught. Eventually, however, a Critical Indigenous Studies approach moves collaborators toward more powerful creative problem solving in and for their communities, which is arguably the end goal of the educational experience. Given access to both Euro-Western and Indigenous meaning-making orientations, collaborators are more likely to overcome barriers of isolation and separation, to contemplate and sort out cultural dissonance, and to purposefully act for the benefit of community. Incorporating Indigenous epistemologies in collaborative contexts develops relationships that are "smooth, steady, and resilient," the three characteristics that Keith Basso (1996) defines as necessary to developing a capacity for wisdom (130–34).

In conclusion, let me put forward a series of quotes I find important to keep in mind. These are recommendations for practice, if you will; and, together, they comprise a measured summary of the ideas behind successfully sharing power within places of intercultural exchange. First, this from Michael Holzman (2003), who says, "Writing well is not merely a matter of writing correctly—it is not simply a [universal] technique. It is a manifestation of participation in a particular civilization." In forwarding this idea, Holzman, a rhetoric and composition scholar, understands what Allen and Crawley teach, which is that all knowledge is contextual and historical, whether it involves rerouting an electrical current or a sentence (1998). What "literacy means," then, needs to transcend our own contexts, our disciplinary approved and regulated practices. This can happen when we meet on middle ground to collaborate.

Second, from Kimmerer (2002): "Intellectual diversity fuels the evolution of cultures and their ability to adapt to a changing world" (434). Remembering Powell's (2004) words, we understand our need to adapt to survive in a world where, increasingly, effective intercultural collaborations undergird any successful venture. Think about the most innovative technologies emerging today; bringing those technologies into being depends on bringing multiple perspectives together for transcultural understanding.[5] Think, for instance, about Kristin Searle and Yasmin Kafai's research introducing electronic textiles (e-textiles) in a Native studies class for American Indian girls (2015). Their research provides one example of adapting culturally responsive "making" for twenty-first-century purposes, or a middle way. Making with e-textiles increases knowledge of computing and engineering practices as it simultaneously increases the power of those who are often marginalized to participate in those practices. Additionally, think about Jennifer Jacobs and Amit Zoran's research into hybrid practice with the Ju/'hoansi in the Kalahari (2015). Their use of 3D digital tools together with hunter-gatherer designs demonstrates how communities gain insight and power through collaborative sojourning. Despite what were initially considered vast differences concerning the worldviews and technology they each brought to the venture, Jacobs and Zoran found the Ju/'hoansi were "engaging in design practices with comparable levels of sophistication" as their own. And they exhibited a keen "awareness of [both] individual style and community norms" (624). Jacobs and Zoran conclude, "*The value of a communal working atmosphere should not be underestimated*" (628, emphasis in original).

Third, but perhaps most important, this from Burkhart (2004): Knowledge is "shaped and guided by human actions, endeavors, desires, and goals" (21). Because of Burkhart, we understand that sojourners must bring their actions, endeavors, desires, and goals to middle ground to achieve alternative images and sought-after outcomes, even contradictory visions of outcomes. This does not necessarily mean changing who we are. It certainly does not mean blaming the other for not being who we want them to be or as we are. Rather, it means acknowledging shared power and knowledge, which are made much more apparent via the relationships collaborators co-create. It means withholding judgment, especially concerning another's literacy practices. And, when we encounter tension, it means we work through it respectfully, responsibly, reciprocally, and relationally. If relationships are conducive to earning and developing trust, they can be sustained over time, which means mutually beneficial work can be performed.

Bringing actions, endeavors, desires, and goals to middle ground means paying attention to what is not happening as well as what is. It necessitates dialogue to learn which questions really matter. It means resetting priorities when needed. It means questioning what is constructed as "legitimate" knowledge. Understandings of what constitutes research data, for example, can be opened up to include experiential narratives, which then brings interpretive force to quantitative measurements. Experience is, as Monroe (2015) argues, a powerful mode of performance (27) and a characteristic that underscores the value of relationship. Additionally, enacting Critical Indigenous Studies approaches means valuing incremental insight arising from small, serendipitous interactions, which includes confronting issues of unequal access to power between sojourners. Confronting power issues leads to greater understanding of how literacy is often systemically normed and regulated. In addition, it allows us to understand how our literate understandings might also be shifted so that literacy practices more easily move between disciplinary discourses, theories, and methodologies. Collaborators accept changes as they adapt to each other and the contexts of their shared environment (Mathison discusses this in the following chapter).

This study of collaborative sojourning suggests approaches not typically in our repertoires, but it argues the benefits and possibility of working in relationships of trust toward shared power. It reimagines sojourning by invoking Indigenous epistemologies, working present and future actions toward greater community benefit. Often described as existing backward and forward along a seven-generational continuum, the scope of such work is long range. It is just one piece of much larger possibilities and does not expect to be definitive. It does, however, encourage small transformations as we continue searching reflexively for answers with more respect, reciprocity, and responsibility toward enduring relationship.

NOTES

1. Since schooling in the United States is based on Euro-Western epistemologies, scholars realized that being socialized differently from mainstream Anglos—particularly via languages other than English—can almost guarantee school failure for certain groups (see Schultz and Hull 2002; see also Jacobs and Jordan 1993). This realization became a call to action for scholars studying and documenting the role of language in learning.
2. The term "Indigenous," as employed by Critical Indigenous Studies scholars, is a theoretical construct used within academic contexts. Its use is not meant to homogenize tribal particularities. Indeed, it acknowledges and honors different tribal traditions, histories, and languages. In asserting Indigenous peoples, value of relatedness, respect, reciprocity, and responsibility, Critical Indigenous Studies does, however, premise a certain commonality across tribal cultures and geographies. Commonality does not imply sameness. Different Native communities may have distinct cultures and traditions, yet maintain the "profound significance" and common valuation of "sovereignty, tradition, and history" (Coffey and Tsosie 2001, 197). These values in turn influence how and why Native peoples come to know, be, and do.

3. Data from the National Science Foundation Science and Engineering Indicators (National Science Foundation 2018) reveal that as of 2015 women still comprised only 25 percent of the science workforce and only 15 percent of the engineering workforce. The numbers are even lower when race is factored in, with Asians constituting 21 percent, Hispanics 6 percent, and Blacks 5 percent (National Science Foundation 2017). Statistics offered by the National Girls Collaborative Project note only 19.3 percent of women were awarded engineering degrees in 2015 (https://ngeproject.org/statistics).
4. Richard White (2011) defines the term middle ground as a process—brought about, in part, by "creative misunderstanding"—from which arises "a set of practices, rituals, offices, and beliefs that although comprised of elements of all the groups in contact, is as a whole separate from the practices and beliefs of all of those groups" (xii–xiii). According to White, it is, in some respects, similar to Levi-Strauss's *bricolage*, where people use accessible "materials" to understand and "overcome obstacle[s]" (White 2011, xiii). White's text focuses on a particular, historical space (*pays d'en haut*), but he notes that middle ground can be (and has been) used as a spatial metaphor, as I have done in this chapter. White notes that middle ground connotes "creation of infrastructure that could support and expand the process" of understanding and overcoming obstacles, and it is only possible when there is "both a rough balance of power and mutual need" (xiii). "Force and violence," he says, "are hardly foreign to the process . . . but *the critical element is mediation*" (xii, emphasis added).
5. According to the Bureau of Labor Statistics, "By 2020 there will be 1.4 million computer-science-related jobs available and only 400,000 computer science graduates with the skills to apply for those jobs. Further, Information Technology (IT) workers have been estimated to earn 74 percent more than the average worker" (https://www.whitehouse.gov/blog/2013/12/11/computer-science-everyone). Additionally, the National Science Foundation's (NSF) goal is to make certain "engaging" and "rigorous academic computer science courses" are placed in "10,000 schools taught by 10,000 well-prepared teachers. Consequently, they developed and implemented two computer science courses—CS Principles . . . and Exploring Computer Science." According to NSF, the courses are designed "to teach the fundamental concepts and big ideas of computing along with coding, and to inspire kids about computer science's creative potential to transform society. These courses were designed to be accessible and engaging for *all* students, with the particular goal of increasing inclusion of women and other groups that are significantly underrepresented in computing." (https://www.whitehouse.gov/blog/2013/12/11/computer-science-everyone).

10
SOJOURNING, RESISTANCE, AND TRUST

Maureen A. Mathison

We have found reflecting together on the discomfort we experienced to be a valuable, knowledge-making process of inquiry. In the end, we suggest that interdisciplinary pedagogy does indeed have the potential to open up exciting opportunities for teaching and learning in higher education. (Friedow et al. 2012, 408)

Interdisciplinary collaboration reveals cultural differences between disciplines as they work side by side, each being exposed to other ways of being and doing (Gee 1996). This certainly was the case in our collaboration with the College of Engineering. Although not intractable, the differences were paradigmatic of the technical sciences and the humanities base worldviews. In the humanities, individual scholarship tends to be common because of the nature of inquiry, and the lone scholar is often more highly valued than those who collaborate. "The literature examining collaboration in the Humanities (particularly), but also in many of the social sciences is much smaller and often predicated simply on showing how little of it there is in comparison with science" (Lewis, Ross and Holden 2012, 694). And while those in the "hard" disciplines collaborate frequently, they generally do so as laboratory partners and coauthors. Working in teams of professors and graduate students they discuss and conduct research, and publish together (Real 2012). Most often, they share the same expertise. In our collaboration, humanities/engineering, the conditions we entered into were distinct from both norms.

Our goal in collaborating was unique in that we determined to utilize both areas of expertise, humanities and the technical sciences, to create something novel out of highly disparate backgrounds.

In their descriptive chapter of a similar collaboration, Erik Fisher and Roop Mahajan (2010) provide an example of how distinct the thought processes of humanities and engineering faculty members are. As part of their collaboration they held regular reading seminars with faculty in both academic areas. Following is an illustration they provide that highlights the differences through the remarks of two professors, one from English and one from engineering:

Department of English
At its best, for me, interpretation creates——recreates—the student as a kind of paranoic. Questioning everything in the world, not just literature but, the hope is beyond literature, the student will be asking questions about the way the world works, including, hopefully, the way that science works, the way that every discipline works, the way in which one perceives the universe to work, and certainly one's own self. (215)

Department of Computer Engineering
My sense is that the best scientists . . . are the ones that are in fact skeptical of themselves, of what they do. Here again, I was most impressed by those articles [in Labinger and Collins 2001] in which there's a kind of assumption that really the greatest and most acute criticisms comes from within the community, and not from without. (215)

The perspective of the humanities professor looks outward; what is learned is anticipated to be of use beyond the immediate subject matter. On the other hand, the technical sciences professor's perspective points inward to an insulated community where what is purposeful can only be utilized in the same context in which it was learned. Juxtaposed, the quotes can also be read as potentially threatening positions to the other in that the humanities indicates its usefulness to those outside its field, while the technical sciences indicate a lack of interest in outsider "help." At its best, collaboration between the two academic areas can be fraught with insignificant misunderstandings that can easily and quickly be dismissed. At its worst, it can be fraught with tensions, where resistance takes time to ameliorate, and trust is slow to develop between collaborators.

As seen in the preceding chapters, initial experiences, more often than not, caused concern or frustration for writing consultants. Throughout the volume "resistance" and "trust" appear, two words critical to understanding collaboration when change occurs on the scale of this program. Resistance to change is common as it "involves going from the known to the unknown" (Bovey and Hede 2001, 372). When the presence of change agents requires change recipients to alter habituated structures and practices, and in this case disciplinary cultures as well, relying on previous set expectations and patterned behaviors becomes counterproductive. As predictability gives way to confusion and uncertainty, participants can no longer fall back on that which has been shown to be effective for them. In our case, writing and engineering were both change agents and change recipients, with each having to receive and enact suggestions for change while also being reciprocally transformed. Writing consultants no longer could rely solely on their disciplinary knowledge and identity, nor could engineers; though change through collaboration is mutual, it requires developing a degree of tolerance for discomfort as innovation inherently destabilizes familiarity.

Change, like other emotion-laden circumstances, has different phases: "(1) denial, (2) resistance, (3) gradual exploration, and (4) eventual commitment" (Bovey and Hede 2001, 372); individuals experience the phases to different degrees and in varying intensities. Emotionality can be equal for both change agents as well as for change recipients as they work through the phases together. Several flash points of change throughout the volume were identified by the writing consultants: (1) the positioning of writing in both disciplines; (2) the subjectivity of graduate students in WID initiatives; (3) the willingness of collaborating partners to learn about the other discipline; and (4) gender issues. Each chapter represents one or more of these flash points, either implicitly or explicitly. Consultants hoped to be listened to, understood, and respected. And they wanted to listen to and understand engineers. From their perspective, they respected the professors and worked creatively with them to establish productive relationships that furthered the goal of the collaboration.

ENGINEERS, HALF OUR TEAM

This volume is written from the writing consultants' perspective, and not that of engineering. Yet, they were half of the collaborative team. To better understand their perception, I contacted ten professors across the College of Engineering who had been very involved in the program and asked if they would be interested in providing their retrospective accounts of the partnership. Seven of the ten professors agreed, with all collaborating departments represented: bioengineering, chemical engineering, civil and environmental engineering, electrical and computer engineering, and mechanical engineering. Two of the seven who agreed had joined the program later, and they provide the perspective of "latecomers." Results are aggregated to maintain anonymity unless permission was given for identification. Depending on their preference, professors were asked the following questions through either interviews or questionnaires:

1. Tell me about the program and pre-program and your involvement with it.
2. What were some of the successes and rewards of the program throughout the years?
3. What were some of the challenges and issues related to the program throughout the years?
4. If you were to design a program like ours in the future, what would you keep from the original model and what would you change?

Interviews took place at the convenience of the professor: over the telephone, in coffee shops, and in offices. They were recorded and transcribed. Other professors preferred answering the questionnaire via email. Responses from both the interviews and questionnaires were parsed according to each of the questions, and using a constant-comparative method three general categories emerged each for affirmative and negative comments (Glaser and Strauss 1967; Corbin and Strauss 2014). These were categorized along the themes of Successes/Rewards and Challenges/Issues (see tables 10.1 and 10.2). Successes/Rewards included positive aspects of teaching and student learning, program impact on outcomes, and interdisciplinary relationships. Challenges/Issues included commentary related to the stumbling blocks of teaching and learning, practical considerations of the implementation of the program, and attitudinal issues, which refers to some of the attitudinal difficulties collaborators experienced.

With factoring (Miles, Huberman, and Saldaña 2014, 286), more descriptive subcategories were generated (see figures 10.1 and 10.2). Factoring involves creating more distinct categories from the larger ones to provide a more fine-grained analysis of the data. These distinct subcategories helped to better understand the experiences of the engineers. Figure 10.1 displays the specific Successes/Rewards subcategories, while figure 10.2 displays the Challenges/Issues subcategories engineering professors mentioned in their interviews and questionnaires.

FROM THE PERSPECTIVE OF ENGINEERS

Successes/Rewards

Professors' positive responses about the program indicated that the collaboration was generally successful, at least in creating a third culture for the engineering students. The resources writing consultants provided for teaching (e.g., handouts, rubrics) supported engineering professors' writing goals for learning. Professors especially appreciated the

Table 10.1. Successes/Rewards of Collaboration

Successes/Rewards General Themes	Operationalization
Teaching and student learning	• Refers to positive comments about instruction and student improvement
Program impact	• Refers to the perceived positive outcomes of instruction
Interdisciplinary relationships	• Refers to the positive interpersonal attitudes toward each other

Table 10.2. Challenges/Issues of Collaboration

Challenges/Issues General Themes	Operationalization
Teaching and learning environment	• Refers to challenging topics about the context of instruction
Practical considerations	• Refers to the challenging day-to-day planning and implementation of collaboration
Attitudinal issues	• Refers to the challenging attitudinal difficulties collaborating parties encountered

Successes/Benefits Specific Themes

Teaching and Student Learning
- Consultants modeled how to teach writing.
- Consultants created resources for writing instruction.
- Consultants created rubrics for feedback and assessment.
- Consultants shared up-to-date information about writing with faculty.
- Consultants provided fresh ways of approaching writing in engineering.

Program Impact
- Students began to value the written text.
- Students appreciated the non-STEM perspective on their writing.
- Student were more prepared for upper-division courses.
- Employers commented that students were more prepared for the workplace.

Interdisciplinary Relationships
- Professors enjoyed the "give and take" of sharing knowledge; they learned about writing from the consultants, as much as consultants learned about engineering from them.
- Professors got to know the consultants and their potential to contribute.

Figure 10.1. A display of "success" subcategories from engineers' responses to research questions.

> *Challenges/Issues Specific Themes*
>
> **Teaching and Learning Circumstances**
> - Consultants had a one- or two-year turnover rate.
> - Consultants entered with a lack of engineering knowledge.
> - Faculty did not work closely with consultants to introduce them to their course goals and to provide a baseline of knowledge to support consultants' teaching.
>
> **Practical Considerations**
> - The amount of writing was considerable given the workload for the degree.
> - The class schedules were difficult to coordinate with consultants' schedules.
> - The time commitment was too little in order for each student to receive feedback and individual attention from the consultants in large classes.
>
> **Attitudinal Issues**
> - Some faculty and students thought writing was a waste of time.
> - There was student resistance to incorporating writing into classes.
> - Some engineers thought their field more rigorous and difficult than the discipline of writing.
> - In some cases, there was little buy-in for the program from engineering faculty.
> - There were student/consultant tensions, particularly in some departments.
> - Engineers sometimes felt the consultants insecure because they were out of field.

Figure 10.2. A display of "challenges" subcategories from engineers' responses to research questions.

instructional support they received from the consultants, many implying they were exposed to implementing novel perspectives (for them) about writing. For example, 30 percent of the respondents claimed they now understood writing in the context of an audience, whereas previously it was not considered. The program also afforded respondents opportunities to see how writing was taught and how students received feedback. Additionally, writing consultants remained current in contemporary writing research, which kept the curriculum competitive for the workplace. The impact of the collaboration left one professor exclaiming, "I still hear the voices of the consultants when I teach now."

Professors also commented that the program impact was positive for student outcomes. Faculty saw that once students had completed earlier courses where writing was taught, they were in the "zone of proximal development," prepared for writing in more advanced courses. Whereas before professors saw students struggle with the writing they were assigned, they now saw students' increased ability to undertake more conceptually difficult assignments. Professors perceived that student writing generally improved. Employers, too, reported to engineering faculty that writing improved. Half of respondents mentioned that employers "anecdotally" perceived students being better prepared. Finally, students also reported to professors that their writing in engineering was enhanced.

A third theme was related to interdisciplinary relationships, best illustrated by one of the biggest champions of our collaboration, a professor of civil and environmental engineering. Professor "R" was highly involved in the program for over ten years, overseeing it from its infancy to its maturation in his department. He collaborated with the writing consultant in the capstone course, mutually developing a class that included the city as a partner. Over the years, they designed the capstone as a feasibility course, where students provided the city with their expertise. By the time students enrolled in the course, he said they had the tools to use writing effectively.

The course was designed so students researched and designed solutions to metropolitan problems and prepared the documents for city engineers, who provided feedback. The course was community engaged in the best sense: students provided a service to the city and city engineers supported students with their feedback, particularly on their writing of the report. According to the professor, students completed over 100 projects over his decade-long supervision. One feasibility report, he said, saved the city approximately $50,000. According to the professor, students also presented their work before the legislative body at the state capitol. "I bought in better than anybody else," he explained when discussing the potential of the program. Professor "R" explained his vision for writing was guided by his decades of experience outside academia. "I got what could be done," he said. In the interview, he also expressed his positive relationship with the writing consultants. He had faith in them, knowing that as he mentored them they would work alongside him. "I was able to be part of their education, too," he explained. "They got to play in a real setting that was good."

Through the years, several consultants collaborated with Professor "R". Each time he learned about the new person, seeking to establish a positive relationship. Today he says, "They're all my friends. I had learned to learn about all these disparate personalities. Where they came from, what their background was, what their nuances were." In many ways, he thought of consultants as colleagues, individuals with whom to establish an open and collaborative environment in which to create together. Several professors commented that successes were due in large part to specific individuals who were invested in the program, such as Professor R. Said one professor, "A lot of things that are innovative are people-dependent."

Challenges/Issues

There were, of course, challenges and issues with the program. Many professors corroborated the concerns and frustrations of the writing consultants with the exception of gender, which was not mentioned, but remains an issue in engineering culture generally, as does the issue of inclusivity (for more on inclusivity, see Beasley and Fischer 2012; Marra et al. 2009; Trenshaw et al. 2013). Professor interviews also provided additional insights into the collaboration that were not considered or that extended some of the challenges consultants mentioned.

The environment for teaching and learning was challenging for writing consultants. They had left their base disciplinary culture and were integrated into a new and unfamiliar one. Five of the seven professors discussed challenges related to consultants' knowledge of engineering. Obviously, as newcomers their knowledge was cursory, and lack of technical knowledge was limiting at times. Respondents commented that consultants did not totally understand engineering concepts, which sometimes made it difficult for graduate consultants to teach students to write as an engineer. "It was hard for the nontechnical consultants to know what was too vague and what was not. . . . it was usually clear it was a miscommunication between disciplines. I could see what they were saying, but honestly, it didn't apply sometimes," explained a professor. Another professor thought that because the consultants did not have engineering backgrounds, "it made it hard to relate to the students and reports" in his class. While writing consultants, because of the immersion context, gradually learned about engineering knowledge and practices, their appointments rarely lasted beyond two years. Turnover meant a new consultant with less knowledge would enter the classroom. Three of the respondents remarked that professors themselves could have done better in making relevant knowledge and practices more explicit to the consultants, shortening the time it took to understand classroom needs and alleviate unnecessary gaps in knowledge.

Practical considerations were also mentioned. At times coordinating schedules was difficult as graduate consultants attended their own courses and fulfilled their own educational commitments, which sometimes conflicted with the needs of the engineering professors. One professor remarked that the time commitment of consultants was limited, and whereas engineering professors and students might work sixty hours a week, consultants did not (although they may have when their own educational commitments were taken into account). And another professor remarked that given the rigorous engineering undergraduate curriculum, it was not practical to assign so much writing to students.

With the exception of one professor, all respondents mentioned attitudinal challenges. They candidly revealed that not all faculty members who were teaching in writing-designated courses were committed or thought it important. "Not everyone saw the value of this training, felt students should just be able to do better if they just tried harder (not true)," revealed a professor. Students sometimes thought the program unreasonable and flawed, as they believed engineering content was more important than the writing component. According to some respondents, students took issue with writing instruction. "Technical aspects somehow are more important and that's what's gonna get me to graduate," said a professor about student attitudes. She continued, students would say to her, "we're wasting our time. I just wanna do my homework." Sometimes the interdisciplinary tension became so great in one particular course that the professor said he had to mediate the consultant-student relationship because of "student hostility." In short, "I think there was a tendency toward mistrust and defensiveness on both ends," he continued. With the dynamics described above, it is no wonder that another professor remarked that "many of the [consultants] felt a little insecure."

Clearly, the program, though successful in many ways, could be improved. Respondents were open about what they would maintain and change from the original. Most professors thought the program model was conceptually and pragmatically solid, but needed refining. Given the comments regarding consultants' lack of technical knowledge, professors asked for more experienced, well-trained consultants with engineers participating in their training. One professor suggested that because engineering professors have a vision for what they need, it would be beneficial to create a list of concepts and terms from which the consultant could learn about the specific discipline and student writing. This professor thought it important for the engineer to work closely with consultants to teach them about their course vision and its associated concepts and terms. Professor "R," who likely had the most positive experience with the program, did just this. He worked closely with the consultants to build a capstone course, allowing them to incorporate writing theory and pedagogy into the curriculum, while teaching them about engineering concepts and practices. He learned from them, too, he said. Another model, where engineering information was shared with the consultant, was implemented into a second department successfully; the consultant ultimately was hired as a full-time employee. "After ten years she is as good as the engineering professors in the types of questions she asks and the comments she provides students as they work on their projects," explained a professor. One professor, however, thought it important to "have the consultants as required resources, but not a consistent presence in the classroom."

Also important to professors was consistency in the quality of instruction. Having consultants assigned to engineering for one or two years was impractical for some. They felt that first-year consultants learned about engineering culture, and by the end of the second year, when they had achieved a more solid understanding of the field, they left and a new consultant entered. This cycle was frustrating for some as seen above.

A second concern was related to issues of respect for writing consultants. One professor identified a dynamic, noting a power disparity between professors and graduate consultants. She exclaimed that professors treated graduate students differently than they would have treated tenure-line professors. Because of this, she proposed creating a joint tenure-line appointment between writing and engineering, hiring someone who could oversee the consultants while also participate in the departments into which they were hired. This would provide more accountability for the program. Students, too, sometimes took issue with graduate consultants, so much so that one professor suggested integrating conflict resolution training into the program. This would help allay tensions as they arose, and perhaps alleviate tensions at their onset. Differences could be aired within a framework for resolution.

Finally, another professor advocated for a different funding model. Once the term of the grant was over the program received ongoing funding from the university, but because of budgetary configurations it was not allocated to both colleges. Previously, the grant had been shared, as the principal investigators represented both engineering and humanities. The program was ultimately discontinued in its original form, eliminating graduate consultants, and in its place three full-time instructors were hired. The decision, according to two respondents, was to hire people who had knowledge of both engineering and writing pedagogy;

full-time instructors would also ensure continuity in instruction over time. Although not asked about the new program, several professors remarked during the interviews that finding qualified people with backgrounds in engineering and writing pedagogy had proven difficult, and so instructors were brought in who did not fully meet the criteria.

SOJOURNING, RESISTANCE, AND TRUST

According to John Berry, Uichol Kim, and Pawel Boski (1988), there are four types of sojourners: (1) an integrator tries to blend both cultures; (2) a marginalizer does not want to be associated with either his [*sic*] home or host culture; (3) a separator seeks to preserve his his/her home culture, but has no interest in establishing relationships within the host culture; and (4) an assimilator downplays or avoids his/her home culture, while embracing the host culture (65).

While the initial program called for an integrative model that blended both cultures, writing consultants initially were at a disadvantage: (1) they possessed little knowledge of engineering culture and practices, (2) they were graduate students, and (3) they were new to a developing collaborative context. However, when professors gave consultants authority to demonstrate their writing expertise in the new environment, together they were able to create a third culture in the classroom in conjunction as they shared each other's knowledge. Integrative sojourning required trust from both partners to fully succeed.

For collaborative success, particularly where partners have different sets of expertise, trust is critical but difficult to initially achieve. Daniel McAllister (1995) defines trust as "the extent to which a person is confident in and willing to act on the basis of the words, actions, and decisions of another" (25). His research on trust has shown two types are important for productive working relationships: cognitive and affective. Cognitive trust is based on a person's performance and reliability. If the person is knowledgeable, performs well, and performs consistently, then he or she will be viewed more positively. Affective trust is based, not on a role associated with work, but on concern and investment in the well-being of others. Cognitive trust, McAlister says, is based on factors extrinsic to the person, while affective trust is based on factors intrinsic to the person (see table 10.3). Essentially, the more knowledge and experience shared, and the more collaborating partners believe in the relationship, the more likely they will succeed in projects.

One challenging issue with these types of trust is that they assume a level of shared knowledge and an ongoing relationship to make a

Table 10.3. Characteristics of Types of Trust

Cognitive Characteristics (extrinsic)	Affective Characteristics (intrinsic)
Based on • knowledge and good reasons; • experience and performance	Based on • concern for individuals; • belief in intrinsic value of relationships; • belief that sentiments are reciprocated

judgment of trustworthiness; both require a basis on which to make it. If there is no prior engagement, then trust must be cultivated. When in-group and out-group members collaborate, trust comes more slowly because difference creates caution, doubt, and even mistrust in some cases. Our writing consultants initially felt more like out-group members than in-group members, making the transition more difficult than had both groups shared a common academic background whereby both types of trust could be developed more quickly.

More recently, research has added another dimension to trust that can further collaboration across cultures for both partners. According to Roy Chua, Michael Morris, and Shira Mor (2012), cultural metacognition mediates trust when people are of different cultural backgrounds. Culling the literature, they define cultural metacognition as the "skill in reflecting on cultural assumptions in order to prepare for, adapt to, and learn from intercultural interactions" (116). In their research, they found a correlation between cultural metacognition and affective trust, but not cognitive trust. "People," they explain, "are more motivated to adjust their schemas if they feel stronger emotional bonds with their partners of different cultures and genuinely want their collaborative relationship to work" (128). Although collaborative relationships can succeed solely with cognitive trust, affective trust allows people to feel more comfortable, and less vulnerable in sharing new ideas (116). This is especially critical to interdisciplinary collaboration where ideas likely are novel to partners representing distinct fields. Additionally, cultural metacognition makes one more mindful of assumptions about others, and upon reflection, one can alter one's behaviors and attitudes moving forward, improving collaborative relations, thus opening up interpersonal space for considering possibilities for innovation.

Chua, Morris, and Mor found that when people working in dyads were of two distinct cultures, understanding and rapport increased when at least one of the persons tried to take the other's perspective (2012, 128), thus increasing affective trust. As Watanabe emphasized in the previous chapter, trust is the bedrock for successful collaboration. Recall Barnhardt's (2002) recounting of the Western European teacher and the Indigenous student both learning from and trusting the other. Each displayed cognitive and affective trust, letting the other take control over decisions at particular moments. This was possible because of their ability to value each other's culture; it afforded them a relationship that benefitted both. This was also supported in Professor "R"'s account of the humanities/engineering collaboration.

In chapter 1 we listed Lim's (2003) five factors that interfere with intercultural interactions: (1) lack of knowledge of the other's culture. (2) ethnocentric assumptions, (3) stereotypes, (4) sociopolitical problems, and (5) belief in universality, and used them as a framework by which to examine our interdisciplinary collaborative experience. Initially many of these factors caused concern on both sides. But in looking at the comments engineers made retrospectively about the program, there is a shift in the discourse about collaboration. Many of Lim's factors became less of an issue once relationships became more established, as consultants noted in their chapters. Many professors (as well as some consultants) transitioned through various phases, from resistance, to gradual exploring and in most cases, eventual commitment (Bovey and Hede 2001, 372). One striking difference was that many faculty members no longer assumed universality about writing, and this was due in part to interactions that helped them better understand disciplinary differences. All professors distinguished between the two disciplines. Over time many of the stereotypes dissipated as professors labored beside the consultants crafting curricula to implement writing. In the interview and questionnaires all remarked the humanities made a positive contribution in students' thinking about writing. In some cases, professors told of their learning about writing through the consultants, maintaining some of the writing concepts, vocabulary, and practices in their classes today.

However, two of Lim's factors persisted—the sociopolitical and ethnocentrism. Respondents talked of professors in their departments who thought writing, though important, was not as demanding or difficult a subject matter as engineering. Students, too (and more explicitly than professors), attitudinally demonstrated ethnocentric views, believing engineering was more central to their careers than was writing. Unfortunately, the program did not change everyone's perspective on the critical relationship between engineering and writing. Some professors and students, in their ethnocentrism, were unable to embrace the critical relevance of writing to engineering success. On a practical level, they observed real improvement in writing, but their biases limited their ability to accept cultural change that would alter perceptions of their own discipline as being writing based, as much as it was technically and scientifically based.

A second issue, and one that directly contributed to the disintegration of the program, was sociopolitical: funding. As explained above, once the grant expired, the funding was allocated to only one of the partnering colleges. This made it difficult for the other to be seen as an equal partner and diminished the ability for joint decision-making. Today, there are no graduate student consultants in engineering, and in a traditionally male-dominated discipline, there are no female writing instructors. In their responses, many engineering professors, without being prompted, lamented the program's demise.

The value of the original model was its collaborative nature, with engineers and writing consultants integrating writing into the curriculum—where it would be learned in the discipline for which it would be used. Combined, the graduate consultants and the engineering professors tell a story of a relatively successful program where members of both disciplines demonstrated an awareness of cultural differences and interest in learning about and from the other, which requires a level of cognitive, affective, and cultural trust. The most productive collaborations involved partners who exhibited the greatest degrees of trust, while the least productive may have had only moderate degrees of cognitive trust, but not an investment in the affective or cultural. In cases such as the latter, instructors might have seen how writing consultants could support their goals, but without interacting much to enhance them through writing. In other cases, professors perceived writing as integral to professionalizing students, but not necessarily in *their* classroom. Or worse, there were those who thought students could learn to write on their own if they "tried harder." Professors who exhibited attitudes such as these likely never made it beyond the stages of denial or resistance to change.

Sojourning reminds us of Ting-Toomey's (2003) call in chapter 1 for mindfulness, where she summons us to create an "awareness of the other person's perspectives, interests, and/or goals, as well as [a] willingness to modify our own interests or goals to adapt" in addressing others from distinct cultures (334). Given that disciplines are comprised of persons with unique cultural characteristics, there are copious situations for potential misunderstandings and tensions between collaborators. Our *Maxims toward a Third Culture*, derived from Lim's factors, are a means to create positive communicative channels to cultivate relationships that supersede disciplinary differences and ultimately promote trust, benefiting collaborative participants. As seen throughout this volume achieving substantial levels of trust is frustrating and time consuming, sometimes easier, sometimes harder. But in cases where both partners are open to trust, collaboration can have an impact.

REFERENCES

ABET Engineering Accreditation Commission. 2018. *Criteria for Accrediting Engineering Programs.* Baltimore, MD: ABET.
Ackerman, John M. 1991. "Reading, Writing, and Knowing: The Role of Disciplinary Knowledge in Comprehension and Composing." *Research in the Teaching of English* 25 (2): 133–78.
Agrawal, Pradeep K. 1997. "Integration of Critical Thinking and Technical Communication into Undergraduate Laboratory Courses." *Proceedings, ASEE Annual Conference,* Milwaukee, WI.
Alber, Mark. 2001. "Creating Writing and Chemistry." *Journal of Chemical Education* 78 (4): 478–90.
Allen, Edward Frank. 1938. *How to Write and Speak Effective English.* New York: Fawcett-Crest.
Allen, Nancy, and Frank Crawley. 1998. "Voices from the Bridge: Worldview Conflicts of Kickapoo Students of Science." *Journal of Research in Science Teaching* 35 (2): 111–32.
Amare, Nicole, and Charlotte Brammar. 2005. "Perceptions of Memo Quality: A Case Study of Engineering Practitioners, Professors, and Students." *Journal of Technical Writing and Communication* 35 (2): 179–90.
Andrews, Deborah C. 2003. "An Interdisciplinary Course in Technical Communication." *Technical Communication* 50 (4): 446–51.
Anson, Chris M., ed. 2002. *The WAC Casebook.* New York: Oxford University Press.
Anson, Chris M., and Deanna Dannels. 2009. "Profiling Programs: Formative Uses of Departmental Consultations in the Assessment of Communication across the Curriculum." *Across the Disciplines,* 6: 1–15.
Archer, Arlene. 2006. "A Multimodal Approach to Academic 'Literacies': Problematizing the Visual/Verbal Divide." *Language and Education* 20 (6): 449–62.
Artemeva, Natasha. 2009. "Stories of Becoming: A Study of Novice Engineers Learning Genres of Their Profession." In *Genre in a Changing World,* edited by Charles Bazerman, Adair Bonini, and Deborah Figueiredo, 158–78. West Lafayette, IN: Parlor Press.
Artemeva, Natasha, Susan Logie, and Jennie St-Martin. 1999. "From Page to Stage: How Theories of Genre and Situated Learning Help Introduce Engineering Students to Discipline-Specific Communication." *Technical Communication Quarterly* 8 (3): 301–16. Accessed October 6, 2018. doi:10.1080/10572259909364670.
Atkinson, Becky. 2008. "Apple Jumper, Teacher Babe, and Bland Uniformer Teachers: Fashioning Feminine Teacher Bodies." *Educational Studies* 44 (2):98–121.
Bachen, Christine M., Moira M. McLoughlin, and Sara S. Garcia. 1999. "Assessing the Role of Gender in College Students' Evaluations of Faculty." *Communication Education* 48 (3): 193–210.
Ball, Jessica. 2010. "Enacting Indigenous Research Ethics through Community-University Partnerships." Accessed October 6, 2018. http://www.ecdip.org/ethics/.
Barnhardt, Ray. 2002. "Domestication of the Ivory Tower: Institutional Adaptation to Cultural Distance." *Anthropology and Education Quarterly* 33 (2): 238–49. Accessed November 28, 2018. https://anthrosource.onlinelibrary.wiley.com/doi/pdf/10.1525/aeq.2002.33.2.238.
Barnhardt Ray, and Angayuqaq Oscar Kawagley. 2005. "Indigenous Knowledge Systems and Alaska Native Ways of Knowing." *Anthropology and Education Quarterly* 36 (1): 8–23. Accessed October 6, 2018. http://dx.doi.org/10.1525/aeq.2005.36.1.008.

Baron, Dennis. 1999. "From Pencils to Pixels: The Stages of Literacy Technologies." In *Passions, Pedagogies, and Twenty-First-Century Technologies*, edited by Gail Hawisher and Cynthia Selfe, 15–33. Logan: Utah State University Press.

Basso, Keith H. 1996. "Wisdom Sits in Places." In *Wisdom Sits in Places: Landscape and Language among the Western Apache*, 105–49. Albuquerque: University of New Mexico Press.

Battiste, Marie. 2002. *Indigenous Knowledge and Pedagogy in First Nations Education: A Literature Review with Recommendations*. Ottawa, ON: National Working Group on Education and the Minister of Indian Affairs. PDF e-book.

Bauer, Henry. 1990. "The Antithesis [[Disciplines as Cultures]]." *Social Epistemology* 4 (2): 215–27.

Bawarshi, Anis. 2003. *Genre and the Invention of the Writer: Reconsidering the Place of Invention in Composition*. Boulder: University Press of Colorado.

Baynham, Mike, and Mastin Prinsloo. 2001. "New Directions in Literacy Research." *Language and Education* 15 (2–3): 83–91.

Bazerman, Charles. 1994. "Systems of Genres and the Enactment of Social Intentions/" In *Genre and the New Rhetoric*, edited by A. Freedman and P. Medway, 79–101. London: Taylor and Francis.

Bazerman, Charles. 2004. Speech Acts, Genres, and Activity Systems: How Texts Organize Activity and People. In *What Writing Does and How It Does It: An Introduction to Analyzing Texts and Textual Practices*, edited by Charles Bazerman and Paul Prior, 309–39. Mahwah, NJ: Erlbaum.

Bazerman, Charles, Joseph Little, L. Bethel, T. Chavkin, D. Fouquette, and Janet Garufis, eds. 2005. *Reference Guide to Writing across the Curriculum*. West Lafayette, IN: Parlor Press.

Beasley, Maya A., and Mary J. Fischer. 2012. "Why They leave: The Impact of Stereotype Threat on the Attrition of Women and Minorities from Science, Math and Engineering Majors." *Social Psychology of Education* 15 (4): 1–22.

Beaufort, Anne. 2012. "Review: The Matter of Writing Assignments in Classes and Beyond." *College English* 74 (5): 477–85.

Becher, Tony. 1981. "Towards a Definition of Disciplinary Cultures." *Studies in Higher Education* 6 (2): 109–22.

Becher, Tony, and Trowler, P. 2001. *Academic Tribes and Territories: Intellectual Enquiry and the Culture of Disciplines*. 2nd ed. Buckingham, England: The Society for Research into Higher Education and Open University Press.

Beer, David, and David McMurrey. 2009. *A Guide to Writing as an Engineer*. 3rd ed. Hoboken, NJ: John Wiley and Sons.

Benyus, Janine. 2005. "Biomimicry's Surprising Lessons from Nature's Engineers." Ted Talk. *TEDTalk*. February 2005. Accessed October 6, 2018. https://www.ted.com/talks/janine_benyus_shares_nature_s_designs?language=en.

Berkenkotter, Carole, and Thomas N. Huckin. 1995. *Genre Knowledge in Disciplinary Communication: Cognition/Culture/Power*. Hillsdale, NJ: Erlbaum.

Berry, John, Uichol Kim, and Pawel Boski. 1988. "Psychological Acculturation of Immigrants." In *Cross Cultural Adaptation: Current Approaches; International Communication Annual*, edited by Young Yun Kim and William B. Gudykunst, 62–89. Newbury Park, CA: Sage.

Berry, Patrick, Gail Hawisher, and Cynthia Selfe. 2012. *Transnational Literate Lives in Digital Times*. Logan, UT: Computers and Composition Digital Press / Utah State University Press. Accessed October 6, 2018. http://ccdigitalpress.org/transnational.

Berthoff, Ann. 1990. *The Sense of Learning*. Portsmouth, NH: Boynton/Cook.

Bix, A. S. 2010. "Engineering National Defense." In *Engineering in a Land-Grant Context*, edited by Alan I. Marcus, 105–33. West Lafayette, IN: Purdue University Press.

Blair, Carole, Julie R. Brown, and Leslie A. Baxter. 1994. "Disciplining the Feminine." *Quarterly Journal of Speech* 80 (4): 383–409.

Booth, Wayne. 1974. *Modern Dogma and the Rhetoric of Assent*. Chicago: Chicago University Press.

Borrego, Maura and Lynita K. Newswander. 2008. "Characteristics of Successful Cross-Disciplinary Engineering Education Collaborations." *Journal of Engineering Education* 97 (April): 123–34.

Borrego, Maura, Lynita Newswander, and Lisa D. McNair. "Applying Theories of Interdisciplinary Collaboration in Research and Teaching Practice." The Thirty Seventh ASEE/IEEE Frontiers in Education Conference. Milwaukee, WI, October 10–13, 2007.

Bovey, Wayne and Andrew Hede. 2001. "Resistance to Organisational Change: The Role of Defence Mechanisms." *Journal of Managerial Psychology* 16 (7): 534–48.

Brammer, Charlotte, Nicole Amare, and Kim Sydow Campbell. 2008. "Culture Shock: Teaching Writing within Interdisciplinary Contact Zones." *Across the Disciplines* 5: 1–17.

Brayboy, Bryan McKinley Jones. 2005. "Toward a Tribal Critical Race Theory in Education." *Urban Review* 37 (5): 425–46.

Brayboy, Bryan McKinley Jones, and Angelina Castagno. 2008. "How Might Native Science Inform 'Informal' Science Learning"? *Cultural Studies of Science Education* 3 (3): 731–50.

Brayboy, Bryan McKinley Jones, Heather R. Gough, Beth Leonard, Roy F. Roehl II, and Jessica Solyom. 2011. "Reclaiming Scholarship: Critical Indigenous Research Methodologies." In *Qualitative Inquiry*, edited by Stephen D. Lapan, Mary Lynn T. Quartoli, and Frances J. Riemer, 423–50. San Francisco: Jossey-Bass.

Brayboy, Bryan McKinley Jones, and Emma Maughan. 2009. "Indigenous Knowledges and the Story of the Bean." *Harvard Educational Review* 79 (1): 1–21.

Britton, James, T. Burgess, N. Martin, A. McLeod, and H. Rosen. 1975. *The Development of Writing Abilities*, 11–18. London: MacMillan Education.

Broad, Bob. 2003. *What We Really Value: Beyond Rubrics in Teaching and Assessing Writing*. Logan: Utah State University Press.

Bromme, Rainer. 2000. "Beyond One's Own Perspective: The Psychology of Cognitive Interdisciplinarity." In *Practising Interdisciplinarity*, edited by Peter Weingart and Nico Stehr, 115–33. Toronto: University of Toronto Press.

Brown, John Seeley, Allan Collins, and Paul Duguid. 1989. "Situated Cognition and the Culture of Learning." *Educational Researcher* 18 (1): 32–42.

Bruffee, Kenneth A. 1984. "Collaborative Learning and the 'Conversation of Mankind.'" *College English* 46 (7): 635–52.

Bucciarelli, Louis L. 1996. *Designing Engineers*. Cambridge, MA: MIT Press.

Buechley, Leah, and Hannah Perner-Wilson. 2012. "Crafting Technology: Reimagining the Processes, Materials, and Cultures of Electronics." *ACM Transactions on Computer-Human Interaction* 19, no. 3 (October): 21, 1–21. Accessed October 6, 2018. doi: 10.1145/1357054.1357123.

Burkhart, Brian. 2004. "What Coyote and Thales Can Teach Us: An Outline of American Indian Epistemology." In *American Indian Thought: Philosophical Essays*, edited by Ann Waters, 15–26. Malden, MA: Blackwell Publishing.

Burnett, Rebecca. 1996. "Some People Weren't Able to Contribute Anything but Their Technical Knowledge: The Anatomy of a Dysfunctional Team." In *Nonacademic Writing: Social Theory and Technology*, edited by Ann H. Duin and Craig J. Hansen, 123–56. Mahwah, NJ: Lawrence Erlbaum.

Byram, Michael. 1997. *Teaching and Assessing Intercultural Communicative Competence*. Clevedon, England: Multilingual Matters.

Cajete, Gregory. 1999. "The Native American Learner and Bicultural Science Education." In *Next Steps*, edited by K. Swisher and J. Tippeconnic, 145–71. Charleston, WV: ERIC Publications.

Cajete, Gregory. 2005. "American Indian Epistemologies." *New Directions for Student Services* 109 (Spring): 69–78. https://doi.org/10.1002/ss.155.

Campbell, Donald. 1969. "Ethnocentrism of Disciplines and the Fish-Scale Model of Omniscience." In *Interdisciplinary Relationships in the Social Sciences*, edited by Muzafer Sherif and Carolyn W. Sherif, 328–48. Chicago: Aldine.

Cardwell, Lea Anna. 2016. "CFP for the WAC Journal." Writing Program Administration Website. February 2. http://www.wpacouncil.org/journal/index.html.

Carli, Linda L. 2001. "Gender and Social Influence." *Journal of Social Issues* 57 (4): 725–41.

Carstensen, Anna-Karin, and Jonte Bernhard. 2009. "Student Learning in an Electric Circuit Theory Course: Critical Aspects and Task Design." *European Journal of Engineering Education* 34 (4): 393–408. Accessed October 6, 2018. doi:10.1080/03043790902990315.

Casmir, Fred L., and Nobleza C. Ascuncion-Lande. 1989. "Intercultural Communication Revisited: Conceptualization, Paradigm Building, and Methodological Approaches." In *Communication Yearbook, 12*, edited by James A. Anderson, 278–309. Newbury Park, CA: Sage.

Chamely-Wiik, Donna M., Jerome F. Haky, and Jeffrey R. Galin. 2012. "From Bhopal to Cold Fusion: A Case-Study Approach to Writing Assignments in Honors General Chemistry." *Journal of Chemical Education* 89 (4): 502–8. Accessed October 6, 2018. doi:10.1021/ed101129v.

Chesler, Mark, and Alford A. Young Jr. 2007. "Faculty Members' Social Identities and Classroom Authority." *New Directions for Teaching and Learning* 111 (Fall): 11–19. doi: 10.1002/tl.281.

Chua, Roy Y. J., Michael W. Morris, and Shira Mor. 2012. "Collaborating across Cultures: Cultural Metacognition and Affect-Based Trust in Creative Collaboration." *Organizational Behavior and Human Decision Processes* 118 (2): 116–31.

Clancey, William. J. 2006. "Observation of Work Practices in Natural Settings." In *The Cambridge Handbook of Expertise and Expert Performance: Its Development, Organization, and Content*, edited by K. Anders Ericsson, Neil Charness, Paul J. Feltovich, and Robert R. Hoffman, 127–45. Cambridge: Cambridge University Press.

Coffey, Wallace and Rebecca R. Tsosie. 2001. "Rethinking the Tribal Sovereignty Doctrine: Cultural Sovereignty and the Collective Future of Indian Nations." *Stanford Law Review* 12 (2): 191–210.

Coleman, S., and Ellen Raider. 2006. "International/intercultural Conflict Resolution Training." In *The SAGE Handbook of Conflict Communication*, edited by John Oetzel and Stella Ting-Toomey, 663–90. Thousand Oaks, CA: Sage Publications.

Collins, Allen, BBN Laboratories, John Seeley Brown, Susan Newman, and Xerox Palo Alto Research Center. 1987. "Cognitive Apprenticeship: Teaching the Craft of Reading, Writing, and Mathematics." *US Department of Education: Office of Educational Research and Improvement Center*. Accessed September 1, 2011. https://eric.ed.gov/?id=ED284181.

Connors, Robert. 1982. "The Rise of Technical Writing Instruction in America." *Journal of Technical Writing and Communication* 12 (4): 329–52.

Connors, Robert. 1997. *Composition-Rhetoric: Backgrounds, Theory, and Pedagogy*. Pittsburgh, PA: University of Pittsburgh Press.

Corbin, Juliet, and Anselm Strauss. 1998. *Basics of Qualitative Research: Techniques and Procedures for Developing Grounded Theory*. Thousand Oaks, CA: Sage.

Cowan, Geoffrey, and Amelia Arsenault. 2008. "Moving from Monologue to Dialogue to Collaboration: The Three Layers of Public Diplomacy." *The Annals of the American Academy of Political and Social Science* 616 (1): 10–30.

Creswell, John W. 2014. *Qualitative Inquiry and Research Design: Choosing among Five Approaches*. Thousand Oaks, CA: Sage.

Cunningham, Don, Jill Stewart, Gabrielle Ness, and Caitlin Webb. 2012. "Perceptions of Professional Engineers and Architects Regarding Effective Technical Communication." *ISRN Education*, 1.

Dannels, Deanna. 2000. "Learning to Be a Professional: Technical Classroom Discourse, Practice, and Professional Identity Construction." *Journal of Business and Technical Communication* 14 (1): 5–37.

Dannels, Deanna. 2003. "Teaching and Learning Design Presentations in Engineering: Contradictions between Academic and Workplace Activities." *Journal of Business and Technical Communication* 17 (2): 139–69.

Dannels, Deanna. 2005. "Leaning in and Letting Go." *Communication Education* 54 (1): 1–5.

Dannels, Deanna, Chris M. Anson, Lisa Bullard, and Steven Peretti. 2003. "Challenges in Learning Communication Skills in Chemical Engineering." *Communication Education* 52 (2): 50–56.

Dannels, Deanna, and Amy L. Housley Gaffney. 2009. "Communication across the Curriculum and in the Disciplines: A Call for Scholarly Cross-Curricular Advocacy." *Communication Education* 58 (1): 124–53.

Darling, Ann. 2011. "Confessions of an Interdisciplinary Teacher in Training." *Spectra* 47 (4): 4–10. https://www.natcom.org/publications/spectra/november-2011.

Deloria, Vine, Jr. 1970. "Power, Sovereignty, and Freedom." In *We Talk, You Listen*, edited by Vine Deloria Jr., 114–37. New York: Macmillan.

Deloria, Vine, Jr. 2001. "Indigenizing Education: Playing to Our Strengths." In *Power and Place*, edited by Vine Deloria Jr. and Daniel Wildcat, 7–19. Golden, CO: Fulcrum Resources.

Devitt, Amy J. 2004. *Writing Genres*. Carbondale: Southern Illinois University Press.

Dillon, Patrick. 2006. "Creativity, Integrativism and a Pedagogy of Connection." *International Journal of Thinking Skills and Creativity* 1 (2): 69–83.

Dillon, Patrick. 2008. "A Pedagogy of Connection and Boundary Crossings: Methodological and Epistemological Transactions in Working across and between Disciplines." *Innovations in Education and Teaching International* 45 (3): 255–62. Accessed October 6, 2018. doi:10.1080/14703290802176121.

Dinitz, Sue, Jack Drake, Shirley Gedeon, Jean Kiedaisch, and Char Mehrtens 1997. "The Odd Couples: Interdisciplinary Team Teaching." *Language and Learning across the Disciplines* 2 (2): 29–42.

Donnell, Jeffrey. 2005. "Illustration and Language in Technical Communication." *Journal of Technical Writing and Communication* 35 (3): 239–71.

Douglas, Elliot P., Mirka Koro-Ljungberg, and Maura Borrego. 2010. "Challenges and Promises of Overcoming Epistemological and Methodological Partiality: Advancing Engineering Education through Acceptance of Diverse Ways of Knowing." *European Journal of Engineering Education* 35 (3): 247–57. Accessed October 6, 2018. doi:10.1080/03043791003703177.

Downs, Doug. 2004 *Teaching Our Own Prison: First Year Composition Curricula and Public Conceptions of Writing*. PhD diss., University of Utah.

Downs, Doug. 2013. What Is First-year Composition? In *A Rhetoric for Writing Program Administrators*, edited by Rita Malenczyk, 50–63. West Lafayette, IN: Parlor Press.

Downs, Douglas, and Elizabeth Wardle. 2007. "Teaching about Writing, Righting Misconceptions: (Re) envisioning 'first-year composition' as 'Introduction to Writing Studies.'" *College Composition and Communication* 58 (4): 552–84.

Eisner, Elliot, and Alan Peshkin. 1990. *Qualitative Inquiry in Education*. New York: Teachers College Press.

Engeström, Y. Yrjö. 1987. *Learning by Expanding: An Activity Theoretical Approach to Developmental Research*. Helsinki: Orienta-Konsultit Oy.

Ericsson, K. Anders. 2006. "An Introduction." *The Cambridge Handbook of Expertise and Expert Performance: Its Development, Organization, and Content*, edited by K. Anders Ericsson, Neil Charness, Paul J. Feltovich, and Robert R. Hoffman, 683–704. Boston: Cambridge University Press.

Eubanks, Philip. 2010. *Metaphor and Writing: Figurative Thought in the Discourse of Written Communication.* London: Cambridge University Press.

Faber, Brenton. 2002. *Community Action and Organizational Change: Image, Narrative, Identity.* Carbondale: Southern Illinois University Press.

Fahnestock, Jeanne. 2003. "Verbal and Visual Parallelism." *Written Communication* 20 (2): 123–52.

Faigley, Lester. 1999. "Material Literacy and Visual Design." In *Rhetorical Bodies: Toward a Material Rhetoric,* edited by Jack Selzer and Sharon Crowley, 171–201. Madison: University of Wisconsin Press.

Fairclough, Norman. 2001. *Language and Power.* Harlow: Pearson Education Limited.

Fairhurst, Gail. T., and Linda Putnam. 2014. "Organizational Discourse Analysis." In *The SAGE Handbook of Organizational Communication: Advances in Theory, Research, and Methods.* 3rd ed., edited by Linda L. Putnam and Dennis K. Mumby, 271–96. Thousand Oaks, CA: Sage.

Faize, Fayyaz Ahmad, Mohammad Arshad Dahar, and Asaf Niwaz. 2010. "Improving Students Performance on Multiple-Choice Questions in Physics: Practice in Justifying the Options." *International Journal of Academic Research* 2 (6): 204–7. http://www.ijar.lit.az/.

Faulkner, Wendy. 2000. "The Power and the Pleasure? A Research Agenda for 'Making Gender Stick' to Engineers." *Science, Technology, and Human Values* 25 (1): 87–119.

Faulkner, Wendy. 2007. "'Nuts and Bolts and People': Gender-Troubled Engineering Identities." *Social Studies of Science* 37 (1): 331–56.

Feld, Steven, and Keith H. Basso, eds. 1996. *Senses of Place.* Santa Fe: School of American Research Press.

Felder, Richard M., and Linda K. Silverman. 1988. "Learning and Teaching Styles in Engineering Education." *Engineering Education* 78 (7): 674–81. Accessed July 2, 2013. http://www4.ncsu.edu/unity/lockers/users/f/felder/public/Papers/LS-1988.pdf.

Fine, Michelle. 2000. *Speed Bumps: A Student-Friendly Guide to Qualitative Research.* New York: Teachers College Press.

Finkelstein, Leo. 2004. *Pocket Book of Technical Writing for Engineers and Scientists.* 2nd ed. New York: McGraw Hill.

Fiore, Stephen M., Robert R. Hoffman, and Eduardo Salas. 2008. "Learning and Performance across Disciplines: An Epilogue for Moving Multidisciplinary Research toward an Interdisciplinary Science of Expertise." *Military Psychology* 20: S155-S170. Accessed October 6, 2018. doi:10.1080/08995600701804939.

Fisher, Erik, and Roop L. Mahajan. 2010. "Embedding the Humanities in Engineering: Art, Dialogue, and a Laboratory." In *Trading Zones and Interactional Expertise,* edited by Michael E. Gorman. Cambridge, MA: MIT Press.

Flower, Linda. 2003. "Talking across Difference: Intercultural Rhetoric and the Search for Situated Knowledge." *College Composition and Communication* 55 (1): 38–68.

Flower, Linda, Eleanor Long, and Lorraine Higgins. 2000. *Learning to Rival: A Literate Practice for Intercultural Inquiry.* Mahwah, NJ: Lawrence Erlbaum Associates.

Ford, Julie Dyke. 2004. "Knowledge Transfer across Disciplines: Tracking Rhetorical Strategies from a Technical Communication Classroom to an Engineering Classroom." *IEEE Transactions on Professional Communication* 47 (4): 301–15.

Fouad, Nadya A., Romila Singh, Mary E. Fitzpatrick, and Jane P. Liu. 2012. *Stemming the Tide: Why Women Leave Engineering.* Milwaukee: University of Wisconsin.

Fraser, Helen, and Andrea C. Schalley. 2009. "Communicating about Communication: Competence as a Factor in the Success of Interdisciplinary Collaboration." *Australian Journal of Linguistics* 29 (1): 135–55.

Friedow, Alison J., Erin E. Blankenship, Jennifer L. Green, and Walter W. Stroup. 2012. "Learning Interdisciplinary Pedagogies." *Pedagogy* 12 (3): 405–24.

Gee, James Paul. 1989. "Literacy, Discourse, and Linguistics: Introduction." *Journal of Education* 171 (1): 5–17.
Gee, James Paul. 1996. *Social Linguistics and Literacies: Ideologies in Discourses.* New York: Falmer Press.
Gee, James Paul. 1999. *An Introduction to Discourse Analysis: Theory and Method.* London: Routledge.
Geisler, Cheryl. 1993. "The Relationship between Language and Design in Mechanical Engineering: Some Preliminary Observations," *Technical Communication* 40 (1): 173–76.
Gianniny, O. Allan, Jr. 2004. "A Century of ASEE and Liberal Education (or How Did We Get Here from There, and Where Does It All Lead?)" In *Liberal Education in Twenty-First Engineering: Responses to ABET 2000,* edited by David S. Ollis, Kathryn A. Neeley, and Heinz C. Langbiehl, 320–46. New York: Peter Lang.
Gider, Franc, Borut Likar, Tomaz Kern, and Damijan Miklavcic. 2012. "Implementation of a Multidisciplinary Professional Skills Course at an Electrical Engineering School." *IEEE Transactions on Education* 55 (3): 332–40. Accessed December 9, 2018. https://ieeexplore.ieee.org/stamp/stamp.jsp?arnumber=6078443.
Glaser, Barney, and Anselm Strauss. 1967. *The Discovery of Grounded Theory.* London: Weidenfeld and Nicholson.
Godfrey, Elizabeth, and Lesley Parker. 2010. "Mapping the Cultural Landscape in Engineering Education." *Journal of Engineering Education* 99 (1): 5–22.
Goggin, Maureen Daly. 1999. "The Tangled Roots of Literature, Speech Communication, Linguistics, Rhetoric/Composition, and Creative Writing: A Selected Bibliography on the History of English Studies." *Rhetoric Society Quarterly* 29 (4): 63–88.
Graber, Glenn C., and Christopher D. Pionke. 2006. "A Team-Taught Interdisciplinary Approach to Engineering Ethics." *Science and Engineering Ethics* 12 (2): 313–20. http://www.springer.com/social+sciences/applied+ethics/journal/11948.
Grant-Davie, Keith. 1997. "Rhetorical Situations and Their Constituents." *Rhetoric Review* 15 (2): 264–79.
Gross, Alan G., Joseph E. Harmon, and Michael S. Reidy. 2002. *Communicating Science: The Scientific Article from the Seventeenth Century to the Present.* New York: Oxford University Press.
Guba, Egon G. 1990. "Subjectivity and Objectivity." In *Qualitative Inquiry in Education,* edited by Elliot Eisner and Alan Peshkin, 74–91. New York: Teachers College Press.
Gudykunst, William B. 1998. *Bridging Differences: Effective Intergroup Communication.* Thousand Oaks, CA: Sage Publications.
Gudykunst, William B., ed. 2003. *Cross-Cultural and Intercultural Communication.* Thousand Oaks, CA: Sage Publications.
Gudykunst, William B., and Young Yun Kim. 1984. *Communicating with Strangers: An Approach to Intercultural Communication.* New York: Random House.
Haas, Angela. 2010. "What Can American Indians Tell Us about Digital and Visual Rhetoric Inquiry?: Forging Intellectual Trade Routes toward a Decolonial Digital and Visual Rhetorics Pedagogy." *Conference on College Composition and Communication.* Louisville, KY, March 2.
Hacker, Sally. 1989. *Pleasure, Power and Technology: Some Tales of Gender, Engineering, and the Cooperative Workplace.* New York: Routledge.
Hagan, Susan M. 2007. "Visual/Verbal Collaboration in Print: Complementary Differences, Necessary Ties, and an Untapped Rhetorical Opportunity." *Written Communication* 24 (1): 49–83.
Hall, Bradford. J. 2005. *Among Cultures: The Challenge of Communication.* 2nd ed. Belmont, CA: Thomas Wadsworth.
Hampden-Turner, Charles M., and Fons Trompenaars. 2000. *Building Cross-Cultural Competence: How to Create Wealth from Conflicting Values.* New Haven, CT: Yale University Press.

Hanitzsch, Thomas. 2007. "Deconstructing Journalism Culture: Towards a Universal Theory." *Communication Theory* 17 (4): 367–85.

Haring-Smith, Tori. 1992. "Changing Students' Attitudes." In *Writing across the Curriculum: A Guide to Developing Programs*, edited by Susan McLeod and Margot Soven, 123–31. Newbury Park, CA: Sage.

Hardy, Cynthia. 2001. "Researching Organizational Discourse." *International Studies of Management and Organization* 31 (3): 25–47.

Haswell, Richard. 1991. *Gaining Ground in College Writing: Tales of Development and Interpretation*. Dallas, TX: Southern Methodist University Press.

Henderson, Kathryn. 1991. "Flexible Sketches and Inflexible Data Bases: Visual Communication, Conscription Devices, and Boundary Objects in Design Engineering." *Science, Technology and Human Values* 16 (4): 448–73.

Herman, Carolyn, Rachel E. Casiday, Roberta K. Deppe, Michelle Gilbertson, William M. Spees, Dewey Holten, and Regina F. Frey. 2005. "Interdisciplinary, Application-Oriented Tutorials: Design, Implementation, and Evaluation." *Journal of Chemical Education* 82 (12): 1871–79. Accessed October 6, 2018. doi:10.1021/ed082p1871.

Herrington, Anne J. [1985] 1994. "Writing in Academic Settings: A Study of the Contexts for Writing in Two College Chemical Engineering Courses." In *Landmark Essays on Writing across the Curriculum*, edited by Charles Bazerman and David R. Russell, 97–124. Davis, CA: Hermagoras Press.

Higgins, Lorraine, Maureen A. Mathison, and Linda Flower. 1992. *The Rival Hypothesis Stance: Thinking and Writing about Open Questions*. Technical Report. Pittsburgh, PA: Mellon Literacy in Science Center, Carnegie Mellon University.

Hill, Catherine, Christianne Corbett, and Andresse St. Rose. 2010. *Why So Few? Women in Science, Technology, Engineering, and Mathematics*. Washington, DC: AAUW.

Hirsch, Penny L., Barbara L. Shwom, Charles Yarnoff, John C. Anderson, David M. Kelso, Gregory B. Olson, and J. Edward Colgate. 2001. "Engineering Design and Communication: The Case for Interdisciplinary Collaboration." *Journal of Engineering Education* 17 (4/5): 343–48.

Hirschfield, Laura Ellen. 2011. *Authority, Expertise, and Impression Management: Gendered Professionalization of Chemists in the Academy*. PhD diss., University of Michigan.

Holzman, Michael. 2003. "Rhetoric/Composition // Academic Institutions / Cultural Studies." *Enculturation* 5 (1). http://enculturation.net/5_1/index51.html.

Howard, Rebecca Moore. 2002. "You Have No Right." In *The WAC Casebook*, edited by Chris M. Anson, 130–32. New York: Oxford University Press.

Howe, Kenneth R. 2009. "Isolating Science from the Humanities: The Third Dogma of Educational Research." *Qualitative Inquiry* 15 (4): 766–84. Accessed October 6, 2018. doi:10.1177/1077800408318302; http://www.eric.ed.gov/PDFS/ED284181.pdf.

Hughes, Brad, and B. Emily Hall 2008. "Rewriting across the Curriculum: Writing Fellows as Agents of Change in WAC." *Across the Disciplines: A Journal of Language, Learning, and Academic Writing* 5. Accessed July 6, 2017. Retrieved from http://wac.colostate.edu/atd/fellows.

Hyland, Ken. 2004. *Disciplinary Discourses: Social Interactions in Academic Writing*. Michigan classics ed. Ann Arbor: University of Michigan Press.

Jablonski, Jeffrey. 2006. *Academic Writing, Consulting and WAC: Methods and Models for Guiding Cross-Curricular Literacy Work*. Cresskill, NJ: Hampton Press.

Jacobs, Jennifer, and Amit Zoran. 2015. "Hybrid Practice in the Kalahari: Design Collaboration through Digital Tools and Hunter-Gatherer Craft.: *Proceedings of Conference on Human Factors in Computing Systems (CHI '15)*, 619–28. Seoul, North Korea: ACM. http://dl.acm.org/citation.cfm?doid=2702123.2702362.

Jandt, Fred. 2007. *An Introduction to Intercultural Communication: Identities in a Global Community*. 5th ed. Thousand Oaks, CA: Sage Publications.

Kanaʻiaupuni, Shawn Malia. 2004. "Kaʻakālai Kū Kanaka: A Call for Strength-Based Approaches from a Native Hawaiian Perspective." *Educational Researcher* 33 (9): 26–32.

Kedrowicz, April. A. 2004. "Negotiating Comfort in Difference: Making the Case for Interdisciplinary Education." *American Society for Engineering Education Conference Proceedings*. Salt Lake City.

Kedrowicz, April, Sundy Watanabe, Damon Hall, and Cynthia Furse. 2005. "Infusing Technical Communication and Teamwork within the ECE Curriculum." *Turkish Journal of Electrical Engineering and Computer Sciences* 14 (1): 41–53.

Keller, Evelyn Fox. 1983. *A Feeling for the Organism: The Life and Work of Barbara McClintock*. New York: W. H. Freeman and Company.

Kim, Young Yun. 2001. *Becoming Intercultural: An Integrative Theory of Communication and Cross-Cultural Adaptation*. Thousand Oaks, CA: Sage Publications.

Kimmerer, Robin. 2002. "Weaving Traditional Ecological Knowledge into Biological Education: A Call to Action." *Bioscience* 52 (5): 432–38.

King, Lisa, Rose Gubele, and Joyce Rain Anderson, eds. 2016. *Survivance, Sovereignty, and Story: Teaching American Indian Rhetorics*. Boulder: Utah State University Press.

King, Thomas. 2003. *The Truth about Stories: A Native Narrative*. Minneapolis: University of Minnesota Press.

Kirkness, Verna J., and Ray Barnhardt. 2001. "First Nations and Higher Education: The Four R's: Respect, Relevance, Reciprocity, Responsibility." In *Knowledge across Cultures: A Contribution to Dialogue among Civilizations*, edited by Ruth Hayoe, and Julia Pan. Accessed October 6, 2018. http://www.ankn.uaf.edu/IEW/winhec/FourRs2ndEd.html.

Klein, Julie Thompson. 1990. *Interdisciplinarity: History, Theory, and Practice*. Detroit: Wayne State University Press.

Klein, Julie Thompson. 2005. "Integrative Learning and Interdisciplinary Studies." *Peer Review* 7 (4): 8–10. Accessed May 6, 2012. http://gallery.carnegiefoundation.org/ilp/uploads/Klein-Interdisciplinary_Studies.pdf.

Klein, Julie Thompson. 2009. "The Rhetoric of Interdisciplinarity." In *The Sage Handbook of Rhetorical Studies*, edited by Andrea Lunsford, Kirt A. Wilson, and Rosa A. Eberly, 265–84/ Los Angeles, CA: Sage.

Kockelmans, J. J. 1979. "Why Interdisciplinarity?" In *Interdisciplinarity and Higher Education*, edited by J. J. Kockelmans, 123–60, University Park: Pennsylvania State University Press.

Kostelnick, Charles. 1996. "Supra-Textual Design: The Visual Rhetoric of Whole Documents." *Technical Communication Quarterly* 5 (1): 9–33.

Kostelnick, Charles, and Michael Hassett. 2003. *Shaping Information: The Rhetoric of Visual Conventions*. Carbondale: Southern Illinois University Press.

Kress, Gunther R., and Theo van Leeuwen. 1996. *Reading Images: The Grammar of Visual Design*. New York: Routledge.

Krometis, Leigh-Anne H., Elena P. Clark, Vincent Gonzalez, and Michelle E. Leslie. 2011. "The 'Death' of Disciplines: Development of a Team-Taught Course to Provide an Interdisciplinary Perspective for First-Year Students." *College Teaching* 59 (2): 73–78. Accessed October 6, 2018. doi:10.1080/87567555.2010.538765.

Kryder, LeeAnne G. 1999. "Mentors, Models, and Clients: Using the Professional Engineering Genres." *IEEE Transactions on Professional Communication* 42 (1): 3–11.

Kumpf, Eric P. 2000. "Visual Metadiscourse: Designing the Considerate Text." *Technical Communication Quarterly* 9 (4): 401–24.

Kuriloff, Peshe C. 1992 "The Writing Consultant: Collaboration and Team Teaching." In *Writing across the Curriculum: A Guide to Developing Programs*, edited by Susan H. McLeod and Margot Soven, 94–108. Newbury Park, CA: Sage.

Labinger, Jay A., Harry Collins. 2010. *The One Culture? A Conversation about Science*. Chicago: University Chicago Press.

Lakoff, George, and Mark Johnson. 2003. *Metaphors We Live By*. 2nd ed. Chicago: University of Chicago Press.
Lamont, Michele. 2009. *How Professors Think: Inside the Curious World of Academic Judgment*. Cambridge, MA: Harvard University Press.
Lardner, Emily, and Gillies Malnarich. 2009. "When Faculty Assess Integrative Learning." *Change: The Magazine of Higher Learning* 41 (5): 29–35. Accessed October 6, 2018. doi:10.3200/CHNG.41.5.28–37.
Lattuca, Lisa R. 2001. *Creating Interdisciplinarity: Interdisciplinary Research and Teaching among College and University Faculty*. Nashville: Vanderbilt University Press.
Lavallée, Lynn. 2009. "Practical Application of an Indigenous Research Framework and Two Qualitative Indigenous Research Methods: Sharing Circles and Anishnaabe Symbol-Based Reflection." *International Journal of Qualitative Methods* 8 (1): 21–40.
Lave, Jean, and Etienne Wenger. 1991. *Situated Learning: Legitimate Peripheral Participation*. New York: Cambridge University Press.
Lee, Suman. 2006. "Somewhere in the Middle: The Measurement of Third Culture." *Journal of Intercultural Communication Research* 35 (3): 253–64.
Lemke, Jay L. 1993. *Talking Science: Language, Learning, and Values*. New York: Ablex.
Leont'ev, Aleksei Nikolaevich. 1981. "The Problem of Activity in Psychology," in *The Concept of Activity in Soviet Psychology*, edited by J. V. Wertsch, 37–71. Armonke, NY: Sharpe.
Lewis, Jenny M., Sandy Ross, and Thomas Holden. 2012. "The How and Why of Academic Collaboration: Disciplinary Difference and Policy Implications." *Higher Education* 64 (5): 693–708.
Liberman, Kenneth. 1999. "From Walkabout to Meditation: Craft and Ethics in Field Inquiry." *Qualitative Inquiry* 5 (1): 4–63.
Lim, Tae-Seop. 2003. "Language and Verbal Communication across Cultures." In *Cross-Cultural and Intercultural Communication*, edited by William B. Gudykunst, 53–72. Thousand Oaks, CA: Sage.
Lincoln, Yvonna S., and Egon G. Guba. 1985. *Naturalistic Inquiry*. Newbury Park, CA: Sage.
Lipka, Jerry. 2002. "Schooling for Self-Determination: Research in the Effects of Including Native Language and Culture in the Schools." Eric Indian Education Research Digest, January 2002. (ERIC Document Reproduction Identification No. ED 459989).
Liu, Shiang-Yao, Chuan-Shun Lin, and Chin Chung Tsai. 2011. "College Students' Scientific Epistemological Views and Thinking Patterns in Socioscientific Decision Making." *Science Education* 95(3): 497–517. Accessed October 6, 2018. doi:10.1002/sce.20422.
Lynch, Michael. 2006. "The Production of Scientific Images: Vision and Re-Vision in the History, Philosophy, and Sociology of Science." In *Visual Cultures of Science: Rethinking Representational Practices in Knowledge Building and Science Communication*, edited by Luc Pauwels, 26–40. Hanover, NH: Dartmouth College Press.
Lyons, Scott. 2010. *X-marks: Native Signatures of Assent*. Minneapolis: University of Minnesota Press.
Lyttle, Allyn, Gina G. Barker, and Terri Lynn Cornwell. 2011. "Adept through Adaptation: Third Culture Individuals' Interpersonal Sensitivity." *International Journal of Intercultural Relations* 35 (5): 686–94.
Macdonell, Diane. 1986. *Theories of Discourse*. Blackwell: Oxford.
Manley, Lucy. 2014. "Embedding the Consultant: A NNES Case Study." PhD diss., Miami University.
Marker, Michael. 2004. "Theories and Disciplines as Sites of Struggle: The Reproduction of Colonial Dominance through the Controlling of Knowledge in the Academy." *Canadian Journal of Native Education* 28 (1–2): 102–10.
Marra, Rose M., Kelly A. Rodgers, Demei Shen, and Barbara Bogue. 2009. "Women Engineering Students and Self-Efficacy: A Multi-Year, Multi-Institution Study of Women Engineering Student Self-Efficacy." *Journal of Engineering Education* 98 (1): 27–38.

Marshall, Catherine, and Gretchen Rossman. 2006. *Designing Qualitative Research*. London: SAGE Publications.

Martin, Judith N., and Michael Hecht. 1994. "Conversational Improvement Strategies for Interethnic Communication: African American and European American Perspectives." *Communication Monographs* 61 (3): 236. Accessed July 8, 2009, from Communication and Mass Media Complete database.

Martin, Judith N., and Thomas K. Nakayama. 1999. "Thinking Dialectically about Culture and Communication." *Communication Theory* 9 (1): 1–25.

Martin Nancy, Pat D'Arcy, Brian Newton, and Robert Parker. [1976] 1994. "The Development of Writing Abilities." In *Landmark Essays on Writing across the Curriculum*, edited by Charles Bazerman and David R. Russell, 33–49. Davis, CA: Hermagoras Press.

Mathison, Maureen. 2000. "'I Don't Have to Argue my Design—The Visual Speaks for Itself': A Case Study of Mediated Activity in an Introductory Mechanical Engineering Course." In *Learning to Argue in Higher Education*, edited by Sally Mitchell and Richard Andrews, 74–84. Portsmouth, NH: Boynton/Cook Heinemann.

McAllister, Daniel J. 1995. "Affect-and-Cognition-Based Trust as Foundations for Interpersonal Cooperation in Organizations." *Academy of Management Journal* 38 (1): 24–59.

McCarthy, Lucille Parkinson. 1987. "Stranger in a Strange Land: A College Student Writing across the Curriculum." *Research in the Teaching of English* 21 (3): 233–65.

McCarthy, Lucille Parkinson, and Stephen M. Fishman. 1991. "Boundary Conversations: Conflicting Ways of Knowing in Philosophy and Interdisciplinary Research." *Research in the Teaching of English* 25 (4): 419–68.

McCloskey, Donald. N. 1985. *The Rhetoric of Economics*. Madison, WI: University of Wisconsin Press.

McDaniel, Elizabeth A., and Guy C. Colarulli. 1997. "Collaborative Teaching in the Face of Productivity Concerns: The Dispersed Team Model." *Innovative Higher Education* 22 (1): 19–36. Accessed October 6, 2018. doi:10.1023/A:1025147408455.

McLeod, Susan H. 1992. "Writing across the Curriculum: An Introduction." In *Writing across the Curriculum: A Guide to Developing Programs*, edited by Susan H. McLeod and Margot Soven, viii. London: Sage.

McLeod, Susan H., and Margot Soven, eds. 1992. *Writing across the Curriculum: A Guide to Developing Programs*. Newbury Park, CA: Sage.

McPeck, John J. 1990. *Teaching Critical Thinking: Dialogue and Dialectic*. New York: Routledge.

MeGill, Allan. 1994. "Introduction: Four Senses of Objectivity." In *RETHINKING OBJECTIVITY*, edited by Allan MeGill, 1–20. Durham, NC: Duke University Press.

Melzer, Dan. 2009. "Writing Assignments across the Curriculum: A National Study of College Writing." *College Composition and Communication* 61 (2): 240–61.

Mertler, Craig A. 2003. *Classroom Assessment: A Practical Guide for Educators*. Los Angeles: Pyrczak Publishing.

Miles, Matthew, and A. Michael Huberman. 1994. *Qualitative Data Analysis: An Expanded Sourcebook*. Thousand Oaks: Sage.

Miller, Carolyn. 1979. "A Humanistic Rationale for Technical Writing." *College English* 40 (6): 610–17.

Miller, Carolyn R. 1984. "Genre as Social Action." *Quarterly Journal of Speech* 70 (2): 151–67.

Miller, Carolyn R. 1994a. "Rhetorical Community: The Cultural Basis of Genre." In *Genre and the New Rhetoric*, edited by Aviva Freedman and Peter Medway, 57–67. London: Taylor and Francis.

Miller, Carolyn R. 1994b. "Genre as Social Action." In *Genre Knowledge and the New Rhetoric*, edited by Aviva Freedman and Peter Medway, 23–42. Bristol, PA: Taylor and Francis.

Miller, Carolyn. 2004. "Reuniting Wisdom and Eloquence within the Engineering Curriculum." In *Liberal Education in Twenty-First Engineering: Responses to ABET 2000*, edited by David S. Ollis, Katherine A. Neeley, and Heinz C. Langbiehl, 41–50. New York: Peter Lang.

Miller, Thaddeus R., Timothy D. Baird, Caitlin M. Littlefield, Gary Kofinas, F. Stuart Chapin III, and Charles L. Redman. 2008. "Epistemological Pluralism: Reorganizing Interdisciplinary Research." *Ecology and Society* 13(2): 1–17.

Mills, Sara. 2004. *Discourse*. New York: Routledge.

Mitchell, Sally. 1992. "Questions and Schooling: Classroom Discourse across the Curriculum." Occasional Paper 1. Hull: University of Hull, Center for Studies in Rhetoric.

Monroe, Barbara. 2015. *Plateau Indian Ways with Words: The Rhetorical Tradition of the Tribes of the Inland Pacific Northwest*. Pittsburgh: University of Pittsburgh Press.

Mullen, Joan. 2008. "Interdisciplinary Work as Professional Development: Changing the Culture of Teaching." *Pedagogy: Critical Approaches to Teaching Literature, Language, Composition, and Culture* 8 (3): 495–508.

Murphy, John D., Alanah J. Davis, and Justin M. Yurkovich. 2009. "Engineering for Interdisciplinary Collaboration." In *E-Collaboration: Concepts, Methodologies, Tools, and Applications*, 1598–1610. Hershey, PA: IGI Global.

National Commission on Writing. 2004. "Writing: A Ticket to Work . . . or a Ticket Out: A Survey of Business Leaders." New York: College Board. Accessed October 6 2018. http://www.collegeboard.com/prod_downloads/writingcom/writing-ticket-to-work.pdf.

National Science Foundation. 2014. "Science and Engineering Indicators 2014." Accessed October 6, 2018. http://www.nsf.gov/statistics/seind14/content/etc/nsb1401.pdf.

National Science Foundation. 2017. "Women, Minorities, and Persons with Disabilities in Science and Engineering." *National Center for Science and Engineering Statistics Directorate for Social, Behavioral, and Economic Sciences*. Accessed October 6, 2018. https://www.nsf.gov/statistics/2017/nsf17310/static/downloads/nsf17310-digest.pdf.

New London Group. 1996. "A Pedagogy of Multiliteracies: Designing Social Futures." *Harvard Educational Review* 66 (1): 60–92.

Newell, W. H. 1994. "Designing Interdisciplinary Courses." *Interdisciplinary Studies Today* (58): 35–51.

Norgaard, Rolf. 1999. "Negotiating Expertise in Disciplinary 'Contact Zones.'" *Language and Learning across the Disciplines* 3 (2): 44–62.

Nowacek, Rebecca S. 2011. *Agents of Integration: Understanding Transfer as a Rhetorical Act*. Carbondale: Southern Illinois University Press.

O'Brien, Alyssa, and Anders Eriksson. 2010. "Cross-Cultural Connections: Intercultural Learning for Global Citizenship." In *Locating Intercultures: Educating for Global Collaboration*, edited by Mara Alagic and Glyn Rimmington, 29–50. New Delhi, India: MacMillan.

Oakeshott, Michael. 1962. *The Voice of Poetry in the Conversation of Mankind*. New York: Basic Books.

Oldenziel, Ruth. 1999. *Making Technology Masculine: Men, Women and Modern Machines in America, 1870–1945*. Amsterdam: Amsterdam University Press.

Paletz, Susannah, Laurel Smith-Doerr, and Itai Vardi. 2010. "National Science Foundation Workshop Report: Interdisciplinary Collaboration in Innovative Science and Engineering Fields."

Palmer, Betsy, and Rose M. Marra. 2004. "College Student Epistemological Perspectives across Knowledge Domains: A Proposed Grounded Theory." *Higher Education* 47 (3): 311–35. Accessed October 6, 2018. doi:10.1023/B:HIGH.0000016445.92289.fl.

Paretti, Marie. 2008. "Teaching Communication in a Capstone Design: The Role of the Instructor in Situated Learning." *Journal of Engineering Education* 97 (4): 491–503.

Paretti, Marie, Lisa McNair, Kelly Belanger, and Diana George. 2009. "Reformist Possibilities? Exploring Writing Program Cross-Campus Partnerships." *WPA: Writing Program Administration–Journal of the Council of Writing Program Administrators* 33 (1–2): 74–113.

Pauwels, Luc. 2006. "A Theoretical Framework for Assessing Visual Representational Practices in Knowledge Building and Science Communications." In *Visual Cultures of Science: Rethinking Representational Practices in Knowledge Building and Science Communication*, edited by Luc Pauwels, 1–25. Hanover, NH: Dartmouth College Press.

Pawley, Alice. 2009. "Universalized Narratives: Patterns in How Faculty Define 'Engineering.'" *Journal of Engineering Education* 98 (4): 309–19.

Penny, Simon. 2009. "Rigorous Interdisciplinary Pedagogy Five Years of ACE." *Convergence: The International Journal of Research into New Media Technologies* 15 (1): 31–54. Accessed October 6, 2018. doi:10.1177/1354856508097017.

Perrin, Daniel, and Marc Wildi. 2010. "Statistical Modeling of Writing Processes." In *Traditions of Writing Research*, edited by Charles Bazerman, Robert Krut, Susan McLeod, Paul Rogers, and Amanda Stansell, 378–93. London: Routledge.

Pettigrew, Karen E. 1999. "Waiting for Chiropody: Contextual Results from an Ethnographic Study of the Information Behaviour among Attendees at Community Clinics." *Information Processing and Management* 35 (6): 801–17.

Piirto, John. 1996. "Teaching Writing to Engineering Students: Toward a Nontechnical Approach." *Journal of Technical Writing and Communication* 26 (3): 307–13. Accessed October 6, 2018. doi:10.2190/232Q-41QQ-JCG5-BYCY.

Plumb, Carolyn, and Richard M. Reis. 2007. "Creating Change in Engineering Education: A Model for Collaboration among Institutions." *Change: The Magazine of Higher Learning* 39(May–June): 22–29.

Poe, Mya, Neal Lerner, and Jennifer Craig. 2010. *Learning to Communicate in Science and Engineering: Case Studies from MIT*. Boston: MIT Press.

Powell, Malea. 2002. "Rhetorics of Survivance: How American Indians Use Writing." *College Composition and Communication* 53 (3): 396–434.

Powell, Malea. 2004. "Down by the River or How Susan La Flesche Picotte Can Teach Us about Alliance as a Practice of Survivance." *College Composition and Communication* 67 (1): 38–60. Accessed October 6, 2018. http://dx.doi.org/10.2307/4140724.

Powell, Malea. 2012. "2012 CCCC Chair's Address: Stories Take Place: A Performance in One Act." *College Composition and Communication* 64 (2): 383–406.

Pratt, Mary Louise. 1991. "Arts of the Contact Zone." *Profession* 91: 33–40.

Prior, Paul. 1998. *Writing/Disciplinarity: A Sociohistoric Account of Literate Activity in the Academy*. Mahwah, NJ: Lawrence Erlbaum.

Prior, Paul. 2004. "Tracing Process: How Texts Come into Being." In *What Writing Does and How It Does It: An Introduction to Analyzing Texts and Textual Practices*, edited by Charles Bazerman and Paul Prior, 167–200. Mahwah, NJ: Earlbaum.

Raider, Ellen, Susan Coleman, and Janet Gerson. 2006. "Teaching Conflict Resolution Skills in a Workshop." In *The Handbook of Conflict Resolution: Theory and Practice*. 2nd ed., edited by Morton Deutsch, Peter Coleman, and Eric C. Marcus, 695–725. San Francisco: Jossey-Bass/Wiley.

Ralston, Patricia, and Cathy Bays. 2010. "Refining a Critical Thinking Rubric for Engineering." Paper presented at the American Association for Engineering Education Annual Conference, Louisville, KY.

Read, Sarah. 2011. "The Negotiation of Writer Identity in Engineering Faculty-Writing Consultant Collaborations." *Journal of Writing Research* 3 (2): 93–117.

Real, Leslie. 2012. "Collaboration in the Sciences and Humanities: A Comparative Phenomenology." *Arts and Humanities in Higher Education* 11 (3): 250–61.

Reave, Laura. 2004. "Technical Communication Instruction in Engineering Schools: A Survey of Top-Ranked U.S. and Canadian Programs." *Journal of Business and Technical Communication* 18 (4): 452–90.

Reese, Dona J., and Mary-Ann Sontag, M. 2001. "Successful Interprofessionial Collaboration on the Hospice Team." *Health and Social Work* 26 (3): 167. Retrieved July 29, 2009, from Academic Search Premier database.

Regaignon, Dara Rossman, and Pamela Bromley. 2011. "What Difference Do Writing Fellows Programs Make?" *The WAC Journal* 22 (November): 41–63.

Riecken, Ted, Frank Conibear, Corrine Michel, John Lyall, Scott Tish, Michelle Tanaka, Suzanne Stewart, Janet Riecken, and Teresa Strong-Wilson, T. 2006. "Resistance through Re-presenting Culture: Aboriginal Student Filmmakers and a Participatory Action Research Project on Health and Wellness." *Canadian Journal of Education* 29 (1): 265–86.

Rogoff, Barbara. 1990. *Apprenticeship in Thinking*. New York: Oxford University Press.

Robinson, John A. 1998. "Engineering Thinking and Rhetoric." *Journal of Engineering Education* 87 (3): 227–29.

Russell, David R. 1995. "Activity Theory and Its Implications for Writing Instruction." In *Reconceiving Writing, Rethinking Writing Instruction*, edited Joseph Petraglia, 51–78. Mahwah, NJ: Earlbaum.

Russell, David. R. 1997. "Rethinking Genre in School and Society: An Activity Theory Analysis." *Written Communication* 14 (4): 504–54.

Russell, David. R. 2002a. *Writing in Academic Disciplines: A Curricular History*. 2nd ed. Carbondale: Southern Illinois University Press.

Russell, David R. 2002b. "Looking beyond the Interface: Activity Theory and Distributed Learning." In *Distributed Learning: Social and Cultural Approaches to Practice*, edited by Mary R. Lea and Kathy Nicoll, 64–82. London: Routledge.

Russell, David R. 2010. "Writing in Multiple Contexts: Vygotskian CHAT Meets the Phenomenology of Genre." In *Traditions of Writing Research*, edited by Charles Bazerman, Robert Krut, Susan McLeod, Paul Rogers, and Amanda Stansell, 353–64. London: Routledge.

Russell, David R., and Patricia Harms. 2010. "Genre, Media and Communicating to Learn in the Disciplines: Vygotsky Development Theory and North American Genre." *Revista Signos* 43 (1): 227–48.

Sageev, Pneena, and Carol J. Romanowski. 2001. "A Message from Recent Engineering Graduates in the Workplace: Results of a Survey on Technical Communication Skills." *Journal of Engineering Education* 9 (October): 685–97.

Savery, John R., and Thomas M. Duffy. 2001. "Problem Based Learning: An Instructional Model and Its Constructivist Framework." Indiana University Center for Research on Learning and Technology, Technical Report No.16–01. Accessed September 9, 2011. http://www.dirkdavis.net/cbu/edu524/resources/Prou7blem%20based%20learning%20An%20instructional%20model%20and%20its%20constructivist%20framework.pdf.

Schultz, Katherine, and Glynda Hull. 2002. "Locating Literacy Theory in Out-of-School Contexts." In *School's Out! Bridging Out-of-School Literacies with Classroom Practice*, edited by Glynda Hull and Katherine Schultz, 11–31. New York: Teachers College Press.

Searle, Kristin, and Yasmin Kafai. 2015. "Culturally Responsive Making with American Indian Girls: Bridging the Identity Gap in Crafting and Computing with Electronic Textiles." *To appear in Proceedings of the Third Conference on Gender IT (ICER'15)*, 9–16. Philadelphia: ACM. Accessed October 6, 2018. http://dl.acm.org/citation.cfm?doid=2807565.2807707.

Semken, Steven. 2005. "Sense of Place and Place-Based Introductory Geoscience Teaching for American Indian and Alaska Native Undergraduates." *Journal of Geoscience Education* 53 (2): 149–57.

Senra, Michael and H. Scott Fogler. 2014. "Teaching Creative Thinking and Transitioning Students to the Workplace in an Academic Setting." *Chemical Engineering Education* 48 (1): 9–16.

Shannon, Claude, and Warren Weaver. 1948. "A Mathematical Theory of Communication." *Bell System Technical Journal* 27 (3): 379–423.

Shapiro, Elayne J., and Carol J. Dempsey. 2008. "Conflict Resolution in Team Teaching: A Case Study in Interdisciplinary Teaching." *College Teaching* 56 (3): 157–62. Accessed October 6, 2018. doi:10.3200/CTCH.56.3.157-162.

Shuman, Larry J., Mary Besterfield-Sacre, and Jack McGourty. 2005. "The ABET 'Professional Skills': Can They Be Taught? Can They Be Assessed?" *Journal of Engineering Education* (January): 685–93.

Smith, Louis. 1990. "Ethics in Qualitative Field Research: An Individual Perspective." In *Qualitative Inquiry in Education*, edited by Elliott Eisner and Alan Peshkin, 258–76. New York: Teachers College Press.

Snow, Charles P. 1959. The Two Cultures. London: Cambridge University Press.

Soliday, Mary. 2011. *Everyday Genres: Writing Assignments across the Disciplines.* Carbondale: Southern Illinois University Press.

Sommers, Nancy. 1980. "Revision Strategies of Student Writers and Experienced Adult Writers." *College Composition and Communication* 31 (4): 378–88.

Sorby, Sheryl A., and William M. Bulleit. 2006. *An Engineer's Guide to Technical Communication.* New York: Pearson.

Spinuzzi, Clay. 2003. *Tracing Genres through Organizations: A Sociocultural Approach to Information Design.* Cambridge, MA: MIT Press.

Star, Susan Leigh, and James R. Griesemer. 1989. "Institutional Ecology, 'Translations' and Boundary Objects: Amateurs and Professionals in Berkeley's Museum of Vertebrate Zoology, 1907–39." *Social Studies of Science* 19 (3): 387–420.

Stephan, Cookie White, and Walter G. Stephan. 2002. "Cognition and Affect in Cross-Cultural Relations." In *The Handbook of International and Intercultural Communication*, edited by William Gudykunst and Bella Mody, 127–42. Thousand Oaks, CA: Sage.

Street, Brian. 1995. *Social Literacies: Critical Approaches to Literacy in Development, Ethnography, and Education.* London: Longman.

Sullivan, Katie, and April A. Kedrowicz. 2012. "Gendered Tensions: Engineering Student's Resistance to Communication Instruction." *Equality, Diversity, and Inclusion: An International Journal* 31 (7): 596–611.

Sullivan, Patricia. 2012. "After the Great War: Utility, Humanities, and Tracings from a Technical Writing Class in the 1920s." *Journal of Business and Technical Communication* 26 (2): 202–28.

Swales, John. 1990. *Genre Analysis: English in Academic and Research Settings.* Cambridge: Cambridge University Press.

Swisher, Karen, and Donna Deyhle. 1989. "The Styles of Learning Are Different, but the Teaching Is Just the Same: Suggestions for Teachers of American Indian Youth." *Journal of American Indian Education* [Special Issue on Learning Styles, August], 1–14.

Taylor, Julie L. 2013. "'What Can You Teach Me?': (Re)thinking Responses to Difference for Multidisciplinary Teamwork." American Society for Engineering Education Conference Proceedings, June 2013, Atlanta, GA.

Terpenny, Janis P., Richard M. Goff, Mitzi R. Vernon, and William R. Green. 2006. "Utilizing Assistive Technology Design Projects and Interdisciplinary Teams to Foster Inquiry and Learning in Engineering Design." *International Journal of Engineering Education* 22 (3): 609–17.

Terry, Thomas M. 1980. "The Narrative Exam: An Approach to Creative Organization of Multiple-Choice Tests." *Journal of College Science Teaching* 9 (3): 156–58.

Thaiss, Chris, and Terry Myers Zawacki. 2006. *Engaged Writers, Dynamic Disciplines: Research on the Academic Writing Life.* Portsmouth, NH: Boynton/Cook HEINEMANN.

Thompson, Jessica Leigh. 2009. "Building Collective Communication Competence in Interdisciplinary Research Teams." *Journal of Applied Communication Research* 37 (3): 278–97.

Ting-Toomey, Stella. 2003. "Teaching Mindful Intercultrual Conflict Management: A Mindful Approach." In *Crossing Cultures: Lessons from Master Teachers*, edited by Nakiye Boyacigiller, Richard Goodman, and Margaret Phillps, 253–68. London: Routledge.

Ting-Toomey, Stella. 2007. "Intercultural Conflict Training: Theory-Practice Approaches and Research Challenges." *Journal of Intercultural Communication Research* 36 (3): 255–71.

Ting-Toomey, Stella, and John G. Oetzel. 2001. *Managing Intercultural Conflict Effectively.* Thousand Oaks, CA: Sage.

Ting-Toomey, Stella, and John G. Oetzel. 2002. "Cross-Cultural Face Concerns and Conflict Styles: Current Status and Future Directions." In *The Handbook of International and Intercultural Communication,* edited by William Gudykunst and Bella Mody, 143–63. Thousand Oaks, CA: Sage.

Tonso, Karen, L. 2007. *On the Outskirts of Engineering: Learning Identity, Gender, and Power via Engineering Practice.* Rotterdam, Netherlands: Sense Publishers.

Toulmin, Stephen. 1958. *The Uses of Argument.* Cambridge, MA: Cambridge University Press.

Tracy, Sarah. 2013. *Qualitative Research Methods: Collecting Evidence, Crafting Analysis, Communicating Impact.* West Sussex, England: Wiley-Blackwell.

Trenshaw, Kathryn F., Ashley Hetrick, Ramona F. Oswald, Sharra L. Vostral, and Michael C. Loui. 2013. "Lesbian, Gay, Bisexual, and Transgender Students in Engineering: Climate and Perceptions." In *Frontiers in Education Conference, 2013 IEEE*: 1238–40.

Tufte, Edward. 1997. *Visual Explanations.* Cheshire, CT: Graphics Press.

Turkle, Sherry, and Seymour Papert. 1990. "Epistemological Pluralism: Styles and Voices Within the Computer Culture." *Signs* 16 (1): 128–57.

Turner, Stephen. 2001. "What Is the Problem with Experts?" *Social Studies of Science* 31 (1): 123–49.

Useem, Ruth Hill. 1999. "Third Culture Kids: Focus of Major Study: TCK Mother Pens History of Field." Retrieved December 9, 2018. http://wwwtck.com/useem/art1.html.

Useem, Ruth H., and Richard D. Downie. 1976. "Third Culture Kids." *Today's Education* (September/October): 103–5.

Vygotsky, Lev S. 1978. *Mind in Society: The Development of Higher Psychological Processes.* Cambridge, MA: Harvard University Press.

Walker, Melanie. 2009. "'Making a World That Is Worth Living': Humanities Teaching and the Formation of Practical Reasoning." *Arts and Humanities in Higher Education* 8 (3): 231–46. Accessed October 6, 2018. doi:10.1177/1474022209339960.

Walvoord, Barbara, and Lucille McCarthy. 1990. *Writing and Thinking in College.* Urbana, IL: NCTE.

Watanabe, Sundy. 2012. "Tensions in Rhetorics of Presence and Performance." PhD diss., University of Utah.

Watanabe, Sundy. 2014. "Critical Storying: Power through Survivance and Rhetorical Sovereignty." In *Crafting Critical Stories: Toward Pedagogies and Methodologies of Collaboration, Inclusion, and Voice,* edited by Judith Flores Carmona and Kristen V. Luschen, 153–70. New York: Peter Lang Publishing.

Webb, Noreen M., Kariane Mari Nemer, and Marsha Ing. 2006. "Small-Group Reflections: Parallels between Teacher Discourse and Student Behavior in Peer-Directed Groups." *Journal of the Learning Sciences* 15 (1): 63–119. Accessed October 6, 2018. doi:10.1207/s15327809jls1501_8.

Weedon, Chris. 1997. *Feminist Practice and Poststructuralist Theory.* 2nd ed. Cambridge: Blackwell Publishers.

Weingart, Peter, and Nico Stehr. 2000. *Practising Interdisciplinarity.* Toronto: University of Toronto Press.

Wertsch, James V. 1994. "The Primacy of Mediated Action in Sociocultural Studies." *Mind, Culture, and Activity* 1 (4): 202–8.

White, Carolyne, Joe Martin, Pat Hays, Guy Senese, Jean Ann Foley, Diane Nuvayouma, and Elaine Riley-Taylor. 2002. "Confronting Tensions in Collaborative Postsecondary Indigenous Education Programs." American Indian and Alaska Native Educational Research Web Site. www.IndianEduResearch.net. ERIC Clearinghouse on Rural Education and Small Schools (ERIC/CRESS). U.S. Department of Education.

White, Richard. 2011. *The Middle Ground: Indians, Empires, and Republics in the Great Lakes Region, 1650–1815.* 2nd ed. New York: Cambridge University Press.

Whitfield, Melinda. 2014. "Negotiating Cultural Encounters." *IEEE Transactions on Professional Communication* 57 (3): 238–39.

Wiggins, Grant, and Jay Tighe. 2005. *Understanding by Design. Alexandria, VA: Association for Supervision and Curriculum Development.* 3rd ed. New York: Wiley & Sons.

Wilson, Shawn. 2001. "What Is an Indigenous Research Methodology?" *Canadian Journal of Native Education* 25 (2): 175–79.

Winsor, Dorothy A. 1990a. "Engineering Writing/Writing Engineering." *College Composition and Communication* 41 (1): 58–70.

Winsor, Dorothy A. 1990b. "Joining the Engineering Community: How Do Novices Learn to Write Like Engineers?" *Technical Communication* 37 (2): 171–72.

Winsor, Dorothy A. 1992. "What Counts as Writing? An Argument Engineers' Practice." *JAC: Journal of Advanced Composition* 12 (2): 33.

Winsor, Dorothy A. 1994. "Invention and Writing in Technical Work: Representing the Object." *Written Communication* 11 (2): 227–50.

Winsor, Dorothy A. 1996. *Writing Like an Engineer: A Rhetorical Education.* Mahwah, NJ: Erlbaum.

Winsor, Dorothy A. 2013. *Writing Like an Engineer: A Rhetorical Education.* London: Routledge.

Wittgenstein, Ludwig. 2001. *Philosophical Investigations.* London: Blackwell.

Wolfe, Joanna. 2009. "How Technical Communication Textbooks Fail Engineering Students." *Technical Communication Quarterly* 18 (4): 351–75.

Wood, Julia T. 2013. *Gendered Lives: Communication, Gender and Culture.* Boston: Cengage Learning.

Wortham, Stanton. 2006. *Learning Identity: The Joint Emergence of Social Identification and Academic Learning.* New York: Cambridge University Press.

Yancey, KathleeN B., Liane Robertson, and Kara Taczak. 2014. *Writing across Contexts: Transfer, Composition, and Sites of Writing.* Logan: Utah State University Press.

Zeleznik, J. M., R. E. Burnett, T. Polito, D. Roberts, and J. Ranks Schafer. 2002. "Roles, and Responsibilities: Crossing the Fine Lines in Cross-Disciplinary Mentorship." *The WAC Casebook*, edited by Chris M. Anson, 227–36. New York: Oxford University Press.

ABOUT THE AUTHORS

LINN K. BEKINS is Associate Professor of Professional Writing and Rhetoric and faculty at the Center for Bio/Pharmaceutical and Biodevice Development at San Diego State University. Her primary research interests focus on the intersection of written communication, learning, and science, and she consults extensively with pharmaceutical, applied science, and healthcare companies interested in improving communication practices between scientists and the general public. She is also extremely active in community-based nonprofit organizations addressing education, environment and sustainability initiatives, and mentors a diverse group of students.

SARAH A. BELL is Assistant Professor of Digital Media at Michigan Technological University. Her research focuses on how emerging technologies mediate the body. She has published in the *Journal of Technical and Business Communication*, *Popular Music in Society*, and *Computational Culture*. Before earning a doctorate, Bell worked for several years as a technical writer.

MARA K. BERKLAND is Professor of Communication at North Central College. Trained in sociolinguistics, her research focuses on effective communication in interdisciplinary classrooms via the socialization of people through language and cultures (both national and disciplinary). She has published in the *Journal on Excellence in Teaching*, and the *Encyclopedia of Women and Islamic Cultures*, as well as in edited volumes. She served as Chair of the Department of Communication at North Central College 2010–16. Berkland was recently awarded her college's Ruge Fellowship for teaching excellence.

DOUG DOWNS is Associate Professor of Writing Studies and former Director of Composition in the Department of English at Montana State University. He researches conceptions and practices of writing, reading, and research among college students, focusing on writing-about-writing pedagogies, undergraduate research, and students' screen-reading habits. Downs is coauthor of *Writing about Writing* (2014) and numerous chapters and articles on WAW instruction, and serves as Editor of *Young Scholars in Writing*, the national peer-reviewed journal of undergraduate scholarship in writing and rhetoric studies.

APRIL A. KEDROWICZ is Assistant Professor of Communication in the Department of Clinical Sciences at NC State University. Kedrowicz's experience includes the development of integrated, discipline-specific communication curricula in engineering and veterinary medicine. Her research interests center around the interplay of communication practices, socialization, gender performance, and professional identity management. Her work has appeared in *Across the Disciplines*, *Communication Education*, *Journal of Business and Technical Communication*, and the *Journal of Veterinary Medical Education*.

MAUREEN A. MATHISON is Associate Professor in the Department of Writing and Rhetoric Studies at the University of Utah. Her research interests focus on tensions in and across disciplines, particularly in the sciences and technical sciences. She teaches courses in the major, including science writing. Her current research examines large-scale controversies in science from an insider perspective. She served as program director of the University Writing Program at the University of Utah for ten years, where she established the Department of Writing and Rhetoric Studies and served as its inaugural chair. She has published in the *Journal of Business and Technical Communication*, *Communication Theory*, *Journal of Literacy*, *Written Communication*, and in numerous edited volumes.

ABOUT THE AUTHORS

SARAH READ is Assistant Professor and Director of Technical and Professional Writing in the English Department at Portland State University. She teaches professional, technical, and scientific writing and rhetoric. Her current research investigates technical documentation and reporting processes at a federally funded supercomputing center for scientific research. In addition, her research develops a writing-about-writing (WAW) approach to teaching professional writing. Her work has appeared in *Technical Communication Quarterly*, *Journal of Business and Technical Communication*, *College Composition and Communication*, *Journal of Writing Research*, and several edited collections.

JULIE L. TAYLOR is an Assistant Professor in the Department of Communication Studies at California State University, San Bernardino. Broadly, her research interests are in organizational communication, gender studies, and interdisciplinary studies. Current research questions investigate organizing of the sex industry, or specifically, asking questions that concern the role of gender in policy construction and implementation, and the consequence of silence as an organizing element of discourse. She has published articles in *Communication Education*, *Connexions: International Professional Communication Journal*, *Journal of Business and Technical Communication*, and the *Western Journal of Communication*.

SUNDY WATANABE, PHD, teaches in the University of Utah's Department of Writing and Rhetoric Studies. Her research advances access and retention for minoritized populations, focusing on intersections of culture, race, rhetoric, and composition within postsecondary institutions. She is particularly interested in the ways educational theories, policies, and practices influence writing praxis. She has published in the journal *Elektrik* and in numerous edited volumes, including *Survivance, Sovereignty, and Story: Teaching Indigenous Rhetorics*.

AUTHOR INDEX

Agrawal, Pradeep, 77
Allen, Edward Frank, 60–61
Allen, Nancy, 157–158, 165
Amare, Nicole, 18, 62
Andrews, Deborah, 25
Anson, Chris, 9, 39
Arsenault, Amelia, 71
Atkinson, Becky, 136

Barnhardt, Ray: on Four Rs, 156–57; on Indigenous education, 159, 186–87
Baron, Dennis, 56
Basso, Keith, 171
Bauer, Henry, 13
Baynham, Mike, 154–55
Bays, Cathy, 77
Beaufort, Anne, 36
Becher, Tony, 22
Beer, David, 123
Benyus, Janine, on storying, 165–67
Bernhard, Jonte, 90–91
Berry, John, 185
Berthoff, Ann, 56
Booth, Wayne, 27
Borrego, Maura, 9, 17
Boski, Pawel, 185
Brammer, Charlotte, 18, 62
Brayboy, Bryan, 165
Broad, Bob, 61
Bromme, Rainer, 30
Bruffee, Kenneth, 16
Bucciarelli, Louis, 25, 107, 111
Buechley, Leah, 162–63
Bulleit, William, 123
Burkhart, Brian, 169, 172
Byram, Michael, 15

Cajete, Gregory, 157, 158, 169
Campbell, Donald, 24
Campbell, Kim Sydow, 18
Cardwell, Lea Anna, 3
Carstensen, Anna-Karin, 90–91
Chua, Roy, 186
Colarulli, Guy, 91
Corbett, Christianne, 29
Cowan, Geoffrey, 71
Craig, Jennifer, 3, 118, 122, 127–28, 132
Crawley, Frank, 157–58, 165
Cunningham, D., 62

Dannels, Deanna, 18, 19, 39
Dashar, Mohammad Arshad, 97
Deloria, Vine, 169
Dinitz, Sue, 14, 23–24
Donnell, Jeffrey, 120
Duffy, Thomas, 94

Eubanks, Philip, 59

Fahnestock, Jeanne, 130–31
Faize, Fayyaz Ahmad, 97–98
Fine, Michelle, 9
Finkelstein, Leo, 123, 130
Fisher, Erik, 175–76
Flower, Linda, 156
Forster, E. M., 16
Fraser, Helen, 22

Gee, James, 57, 68
Gianniny, O. Allan, 72
Godfrey, Elizabeth, 137, 143
Goggin, Maureen, 60
Grant-Davie, Keith, 56
Griesemer, James, 22–23
Gudykunst, William, 14

Hacker, Sally, 144
Hagan, Susan, 130
Hall, Emily, 6
Haring-Smith, Tori, 6–7
Harms, Patricia, 48
Haswell, Richard, 59, 61
Hill, Catherine, 29
Hirschfield, Laura Ellen, 28
Hirsh, Penny, 93
Holzman, Michael, 171
Hughes, Brad, 6
Hull, Glynda, 154

Jablonski, Jeffrey, 16–17, 19
Jacobs, Jennifer, 172
Johnson, Mark, 59

Kafai, Yasmin, 172
Kawagley, Angayuqaq (Oscar), 156–57
Kedrowicz, April, 135, 136, 137
Kim, Uichol, 185
Kimmerer, Robin, 167–68, 171
Kockelmans, J. J., 24

AUTHOR INDEX

Kostelnick, Charles, 122
Kuriloff, Peshe, 86

Lakoff, George, 59
Lamont, Michele, 26
La Place, Pierre Simon, 106
Lardner, Emily, 92
Lattuca, Lisa, 15
Lavallée, Lynn, 162
Lee, Suman, 35
Lerner, Neil, 3, 118, 127–28, 132
Liberman, Kenneth, 160
Lim, Tae-Seop, 20
Lynch, Michael, 120

Mahajan, Roop, 175–76
Malnarich, Gillies, 92
Marker, Michael, 156
Martin, Judith, 32
McAllister, Daniel, 185
McClintock, Barbara, 18
McCloskey, Donald, 26–27
McDaniel, Elizabeth, 91
McLeod, Susan, 16, 71
McMurrey, David, 123
McNair, Lisa, 9
McPeck, John, 75
Melzer, Dan, 37
Mertler, Craig A., 97
Meyer, Manulani, 156
Miller, Carolyn, 6, 37
Miller, Thaddeus, 90
Monroe, Barbara, 157
Mor, Shira, 186
Morris, Michael, 186

Nakayama, Thomas, 32
Newell, W. H., 22
New London Group
Newswander, Lynita, 9, 17
Niwaz, Asaf, 97–98
Norgaard, Rolf, 18
Nowacek, Rebecca, 48, 49

Papert, Seymour, 162, 163
Paretti, Marie, 5, 17
Parker, Lesley, 137, 143
Perner-Wilson, Hannah, 162–63
Perrin, Daniel, 56, 58–59
Poe, Myra, 3, 118, 122, 127–28, 132
Pratt, Mary Louise, 18

Prinsloo, Mastin, 154–55
Prior, Paul, 56

Ralston, Patricia, 77
Ramus, Peter, 60
Reese, Dona, 27
Robinson, John, 120
Russell, David, 37, 48

St. Rose, Andresse, 29
Savery, John, 94
Schalley, Andrea, 22
Schultz, Katherine, 154
Searle, Kristin, 172
Semken, Steven, 157, 158–59
Snow, Charles, 88–89
Soliday, Mary, 7
Sontag, Mary-Ann, 27
Sorby, Sheryl, 123
Spinuzzi, Clay, 48
Star, Susan, 22–23
Stehr, Nico, 28
Sullivan, Katie, 135, 136, 137
Sullivan, Patricia, 24–25

Taylor, Julie, 137
Terry, Thomas, 98
Thaiss, Chris, 8
Thompson, Jessica Leigh, 21
Ting-Toomey, Stella, 33–34
Toulmin, Stephen, 106
Trowler, Paul, 22
Tufte, Edward, 132; on visual discourse, 121–22, 129
Turkle, Sherry, 162, 163

Vinci, Leonardo da, 121, 133

Wagner, Michael, 22
Walker, Eric, 29
Weingart, Peter, 28
Wertsch, James, 109
Whitfield, Melinda, 10
Wildi, Marc, 56, 58–59
Winsor, Dorothy, 23, 45, 50, 61, 121, 131
Wolfe, Joanna, 118, 121

Zawacki, Terry M., 8
Zeleznik, J. M., 18
Zoran, Amit, 172

SUBJECT INDEX

AAAs. *See* American Association for the Advancement of Science
AAUW. *See* American Association of University Women
ABET. *See* Accreditation Board for Engineering and Technology
ABET 2000, 6
academics: assumed universality by, 20–21; ethnocentrism in, 24–26; stereotypes in, 26–27
Academic Writing Consulting and WAC (Jablonski), 16–17
Accreditation Board for Engineering and Technology (ABET), 77, 138, 153(n1)
Across the Disciplines (Hughes and Hall), 6
activity theory, 10
agents of integration approach, 48
allatonceness, 56, 63
American Association for the Advancement of Science (AAAs), 11, 12
American Association of University Women (AAUW), on gender and STEM, 29–30
analysis, 88; in interdisciplinary teaching, 97–98; rhetorical, 119–20
application, 88; in mechatronics class, 94–96
apprenticeships, 96–97, 131
argument, in mechanical engineering, 105–8, 115–17
assessment, writing, 68–69
assignments, 36–37, 92, 93; laboratory memo, 100–104; writing to communicate, 79–80, 83–85; writing-to-learn, 78–79, 81–83
assimilation, 155, 185
assumptions, 67
attitudes, challenges and issues, 180(fig.), 183
audiences, 8, 82; writing for, 95, 119; and writing styles, 79, 80
authenticity, in instruction, 142–43
authority, 28, 155, 162

belief systems, 88
biology, and engineering, 165–66
biomimicry, 166
bioprocessing, 166
boundary crossing, disciplinary, 168

boundary object, writing as, 23
bridging, 70
Brown University, writing fellow programs, 6–7

challenges, in collaboration, 179(table), 180(fig.), 182–85
chart junk, 129
chemical engineering, courses in, 40–42, 46
Chemical Engineering Undergraduate Curriculum Committee, 86
city engineers, 181
civil engineering, writing in, 51–55
clarity, 119
classroom, 109; gender dynamics of, 136–37, 146–50
collaboration, 10, 14, 15, 16, 17, 33, 68, 71, 82, 87, 131–32, 157, 170, 172, 175–76, 186; challenges and issues in, 182–85; compromise and, 31–32; in Critical Indigenous Studies approach, 156, 157–61; CUNY program, 7–8; gender issues in, 150–52; hierarchy in, 164–65; and intercultural dissonance, 161–64; interdisciplinary, 3–4, 9–10, 12, 18, 140–41, 150–52; in learning, 165–67, 177; middle ground, 168, 174(n4); negotiation, 32, 63–64; successes and rewards in, 178–82
"Collaborative Learning and the 'Conversation of Mankind,'" 16
collaborative learning theory, 16
collaborative negotiation model, 31–32
College of Engineering, 4, 7, 139, 175; collaboration in, 177–78; instructional activities, 66–68; instructional emphasis, 65–66; writing in, 39, 36–37, 42–48
College of Humanities, 4
colonialism, and Indigenous learning, 167–68
commitment, reciprocity of, 19
common ground, diplomacy in, 76–77
common sense, 144
communication, 6, 21, 108, 133; civil engineering, 54–55; and disciplinary knowledge, 41–42; drawing as, 113–14; in engineering courses, 138, 139–40; engineering design, 111–13; as feminine,

SUBJECT INDEX

134, 144–45; gendered, 136–37; and innovation, 165–66; intercultural, 10, 14; in materials science and engineering, 42–43; through storying, 165–66; utility of, 141–42; visual, 118–19, 121; writing assignments, 83–85
communities, and collaborators, 157
competence, 138; intercultural, 15–16
composition, 58, 106, 107
compromise, and collaboration, 31–32
conflicts, dialog and, 30–31
consultants, 5, 7, 10, 11, 12, 15–16, 54–55, 151–52; challenges to, 182–84; expertise of, 63–65; female, 149–50
consulting model, 17
contact zones, 18
content, 68, 144; engineering, 39, 64–65
context(s), 67; writing of, 56–57, 64–65
contextualization, 90
contingency, 57
contract, civil engineering, 53
correspondence, 53
course evaluation, gendered discourse in, 139–41
courses, 39, 105; chemical engineering, 40–42; civil engineering, 54–55; integrative teaching, 93–99; interdisciplinary, 87–88
creativity, 33
Critical Indigenous Studies, 167–68, 170, 171, 173; collaboration in, 157–61; Four Rs in, 156–57
critical thinking, 74–76, 80, 91; as learning goal, 77–78
Cross-Cultural and Intercultural Communication (Gudykunst), 14
cross-cultural writing, 62; consultant expertise, 63–65; instructional activities, 66–68; instructional emphasis, 65–66; writing assessment, 68–69
cultures, 72, 137, 146, 171, 186, 187; disciplinary, 13–14, 22–24, 50–51, 135
culture shock, 18
CUNY, 7–8
curriculum, 4, 5, 7, 8, 11, 29; content and syntax, 63–64; critical thinking in, 74–76; Department of Chemical Engineering, 78–81; resistance in, 73–74
curriculum development, 33; civil engineering, 54–55

data, 118, 132
data graphics, 123, 124, 127–29
data tables, 123, 124, 127–29
Department of Chemical Engineering, 11, 71; critical thinking, 74–76; curriculum model, 78–81; teaching and learning objectives, 76–78; sample assignments, 81–85
Department of Civil Engineering, 51
Department of Energy, 52
Department of Mechanical Engineering, 4
Department of Transportation, 52
description, in materials science and engineering, 42
design, 53; in mechanical engineering, 111–14, 116–17
design courses: argument in, 115–17; communication in, 111–14; writing and speaking in, 109–10; zones of proximal development in, 4–5
design practices, 172
design process, 25; writing in, 114–15
design projects, team, 110
diagrams, 123, 126–27
dialectics, 60
dialogs, conflict and, 30–31
diplomacy, 11, 73, 85; common ground in, 76–77
disciplinary conventions, genre and, 43
disciplines, 48, 75, 151, 153, 176; cultural concepts of, 50–51; cultures of, 13–14, 22–24, 135; stereotypes of, 26–28
discourse(s), 21, 23, 36, 47, 57, 68; gendered, 135–37, 139–41; masculine, 137–38, 149; visual, 121–22
discourse communities, 8
dissonance, intercultural, 161–65
diversity, intellectual, 171
documentation, 73; in civil engineering, 53–54; professional, 83–84
drawing, in mechanical engineering, 113–14

EC 2000, 6
editing, competence in, 144
education, Euro-Western approaches, 169
educational development program, 159
electrical engineering class: notebook, 47–48; writing in, 43–45
embodiment, 57
emotionality, 177
empirical method, 163
Engaged Writing, Dynamic Disciplines: Research on the Academic Writing Life (Thaiss and Zawacki), 8
engineering, 3, 7, 50, 89, 174(n3); and biology, 165–66; communication, 111–14; disciplinary culture, 22, 176; female/feminine in, 146–50; gender issues in, 29–30, 134, 135, 136; graphical elements in, 122–31; learning

environments, 90–91; as male dominated, 29–30; masculine discourses in, 137–38; rhetoric in, 116–17; writing in, 38, 48–49, 163–64
engineering reports, 53
Engineering Schools of the West Initiative (ESWI), 6
engineers, 50; successes and rewards in collaboration, 178–82
environmental engineering, 67
Environmental Protection Agency, 52
epistemic writing, 50, 58, 60
epistemologies, 173(n1); adoption of, 162–63; Indigenous, 161; shared, 88, 157–58
ESWI. *See* Engineering Schools of the West Initiative
ethnicity, issues relating to, 11–12
ethnocentrism, 24–26, 134, 151, 152–53, 187; of students, 143–44
ethnographic observation, 4
ethnography, knowledge in, 160
Ethnography of Communication theory, 154
Euro-Western worldview, 155, 156, 158, 160, 169, 173(n1)
evaluation(s), 119; gendered discourse in, 139–41; writing-to-communicate, 80, 84–85; writing-to-learn, 82–83
Everyday Genres: Writing Assignments across the Disciplines (Soliday), 7
exams, interdisciplinary narrative, 97–99
exigence, of writing, 56–57
experiential learning, 46–47
experiments, experimentation: learning through, 161–62; writing assignments and, 41, 43, 44–45, 47, 95
expertise, 28, 59, 64
expressive writing, 79, 80

faculty, 138, 176; on challenges and issues, 182–84; civil engineering, 54–55; collaboration with, 177–78; on critical thinking, 75–76; feedback, 68, 69; on successes and benefits, 178–82; and value hierarchies, 164–65
feasibility studies/reports, 53
Federal Aviation Administration, 52
feedback, 181; peer, 144–45; student, 12, 68–69, 82–83
females: power dynamics, 136–37; sexualized, 147–49
feminine: communication as, 136, 144–45; disciplining, 149–50
field reports, 54
figures, 124

first-year composition courses, 105
Fog Index, 46
form, and content, 144
formal reports, 44–45, 48, 73
Four Rs, in Critical Indigenous Studies approach, 156–57
funding, program, 188

gender, 134, 135, 188; classroom dynamics, 136–37, 146–50; in communication-engineering collaboration, 150–51; hard and soft disciplines and, 11–12; sociopolitics of, 28–29; stereotypes of, 29–30
genres, 57, 123, 131, 145; in civil engineering, 53–54; as instructional tools, 5, 37, 43
goals, and perspectives, 22
grading, 61
graduate students, 5–6, 7, 72
grammar, subject-matter expertise and, 64
graphical elements, rhetorical contexts of, 119–21, 122–31
graphomotoric functions, 58–59
guidelines, 25
guides, to disciplinary knowledge, 36–37

hardware/software model, 59
Hewlett Foundation Grant, William and Flora, 6
hierarchies: in collaboration, 164–65; gender, 151
higher-order concerns (HOCs), 84
How to Write and Speak Effective English (Allen), 60–61
humanities, 90, 91, 134, 162; disciplinary culture, 22, 176; and sciences, 88–89, 99–100

ideas, organizing and developing, 119
identities, 134, 136
illustrations, 123; rhetorical analysis of, 130–31
inclusivity, 12
indigenous peoples, 158, 171, 173(n2), 187; colonialism and, 167–68
information design, 132
infrastructure, and civil engineering, 52
innovation, communication in, 165–66
inscription, 58, 59, 60–61, 65
instruction: activities in, 66–68; authenticity, 142–43; emphasis in, 65–66; and practice, 132–33; resistance to, 143–46; utility of, 141–42; writing, 69–70
instructors: gender dynamics and, 136–37; intercultural dynamics, 163–64

SUBJECT INDEX

integration, 3, 185; of communication, 139–40; of instruction and practice, 132–33
interaction, 108, 187
intercultural communication, 10, 14, 20
intercultural competence, 15–16
interdisciplinarity, 3–4, 9–10, 12, 18, 92, 93; and ethnocentrism, 24–26; gender relations in, 150–51; reciprocity in, 34–35; sociopolitics of, 28–29; and stereotypes, 26–28
interpretation, 50
instructional resources, 41(table)
interviews, 39–40
issues, in collaboration, 179(table), 180(fig.), 182–85

Ju/'hoansi, 172

knowledge(s), 34, 59, 87, 90, 109, 155, 160, 187; disciplinary, 15, 22–24, 36–37, 41–42, 43, 48; gatekeeping vs. mentoring, 46–47; Indigenous, 158–59; inscription of, 60–61; legitimate, 172–73; relational, 169–70; rhetorical, 27–28; shared, 35, 185–86; traditional assumptions of, 148–49; worldview theory, 157–58

labeling, gender dynamics, 149
laboratory course, 110; mechanical engineering, 93–99
laboratory reports, 73, 116, 170; critical thinking in, 74–75
language, 8, 60, 88; and social structures, 135–36
leaning in, 18–19
learning, 12, 89, 152, 158, 177, 179(fig.); biological and engineering, 166–67; critical thinking process in, 75, 77–78; department objectives, 76–77; experiential, 46–47; Indigenous modes of, 158–59; integrated, 87–88, 92; intercultural, 159–60; and pedagogy, 161–62; situated, 4–5, 7; and writing, 36, 39–40, 42–43, 78–79, 81–83
learning environment, engineering, 90–91
"Learning Objectives for Communication and Teamwork in ChE," 86
Learning to Communicate in Science and Engineering: Case Studies from MIT (Poe, Lerner, and Craig), 3
lecture, 110
literacy, literacy practices, 131, 154–55, 160, 171
literature review, 47
lower-order concerns (LOCs), 84

macrodiscourses, 135
marginality, 22–23
marginalizing, 185
masculinity, masculine, 136, 149, 151
materiality, of writing, 57
materials science and engineering class, writing in, 42–43, 46–47
mathematics, technical thinking, 107
meaning making, 171
mechanical engineering, 25, 107, 110; argument in, 106–8; drawing in, 113–14; integrated course in, 93–99; laboratory memo assignment, 100–104; writing, 105–6; writing assignments, 118–19
mechatronics: analysis, 97–99; application, 94–96; course planning, 93–94; mirroring, 96–97
mediation, of disciplinary knowledge, 37
memos, 84, 119; laboratory, 100–104
mentoring, interdisciplinary, 19
merging, 69–70
metacognition, cultural, 186
metadiscourse, and graphical elements, 125–26
metaphors, 67
microdiscourses, 135
middle ground, 168, 174(n4)
mindfulness, 33
mirroring, 88, 96–97
miscommunication, 11, 14, 137
MIT, visual presentation, 118
motives, for writing, 57
Museum of Vertebrate Zoology, 23

narrative style, 10
National Science Foundation (NSF), inclusivity, 11–12
negotiation, collaborative, 31–33, 63–64
New Literacy and Composition Studies, 156
New Literacy Studies (NLS), 154
New London Group, 34
NLS. *See* New Literacy Studies
nontextual material, 118
notebooks, electrical engineering class, 44, 45, 47–48
NSF. *See* National Science Foundation

object, writing as, 46
objectives, teaching and learning, 76–77
objectivity, 162
object world, 25, 107
Ohm's Law, 106
"Only Connect," 16
ontology, shared, 157–58
organicism, 63
other, knowledge of, 22–24

parallelism, visual, 130–31
pedagogy: and learning, 161–62; of visual communication, 131–33
peer review, 84, 85
performance, 36
perspective, 22
Philosophical Essay on Probabilities (La Place), 106
photos, 130
physics, technical thinking, 107
place, Indigenous sense of, 159
power: and gender relations, 12, 149–50; and sociopolitical status, 28–29
practice, instruction and, 132–33
preengineering students, 109–10
presentations, civil engineering, 52
Pro-Engineer, 111
professional development, communication in, 142
professional documents, writing-to-communicate assignments, 83–84
professional experience, learning, 44
professionalism, and female body, 148
Professional Practices class, 64, 65, 66
proofreading, 64
proposals, 53, 116, 119
purpose, 119

qualifications documents, 53
questionnaires, to engineering faculty, 177–78

race, 174(n3); issues relating to, 11–12
reading seminars, 176
real-life experiences, 10
reciprocity, 14, 156; in collaborative learning, 160, 167; in interdisciplinarity, 34–35
reflection logs, 23
regression, in syntax, 59
regulation, and literacy practices, 155
relationship(s), 138, 156, 158, 179(fig.)
reports, 53, 54, 80, 119; formal, 44–45, 48, 73; laboratory, 73, 96
request for proposals, 52–53
research, in civil engineering, 52
research papers, rhetorical analysis of, 119–20
resistance, 72, 107, 176–77; in curriculum structure, 73–74; to female instructors, 136–37; to instruction, 143–46, 164
respect, 156, 158, 184
responsibility, 156
revision, 58
rewards, in collaboration, 178–82
rhetoric(s), 8, 107; in engineering design, 116–17; hard vs. soft science, 162–63; of writing, 56–58

rhetorical analysis, of graphical elements, 119–21, 122–31
rhetorical instruction, 45
rhetorical knowledge, stereotyping and, 27–28
rhetorical writing, 10–11, 36, 56–58, 64
robotics, mechatronics, 95–96
rules of discourse, 21

scaffolding, 4
scaling, 166
schematics, 123, 126–27
scholars: assumptions of universality, 20–21; intercultural, 30–31
science, 174(n3); and humanities, 88–89, 99–100; and Indigenous modes of knowledge, 158–59; shared epistemology and ontology, 157–58; soft vs. hard, 162–63
science-humanities divide, 88–89
scribal writing, 10–11, 36, 50, 63, 65; concepts of, 58–62
segregation, science and humanities, 89
sense of place, Indigenous, 159
separation, 185
sexualization, of female body, 147–49
Shannon-Weaver Communication Model, 46
sketching, technical, 112–13
socialization, 87, 142, 146, 155; in interdisciplinary teaching, 151–52
social organization, of gender, 135
sociocultural patterns, carriers of, 109
sociopolitics, 187, 188; gender issues, 28–29
sojourners, sojourning, 15, 34–35, 163, 185–88; collaborative, 160, 167–68, 172, 173
SolidWorks, 131
specialization, writing in engineering as, 47
specifications, civil engineering, 52, 54
speed bumps, 9–10
stakeholders, civil engineering, 52
stance of inquiry, 46
statistics, literacy in, 132
STEM (science, technology, engineering, and math disciplines), 134; gender issues in, 28, 29–30
stereotypes, stereotyping, 32, 152–53, 187; disciplines of, 26–28; gender, 29–30, 134, 149–50
storying, communication through, 165–67
storytelling, 10
students, 5, 17, 23, 89, 134; civil engineering, 53–54; and communication

consultants, 152–53; design courses, 109–10, 114–15; engineering, 137–38; evaluation by, 68–69, 81–85, 141–46; gender dynamics, 12, 146–50; in-group and out-group, 153, 186; successes and benefits to, 179(fig.), 180–81; use of argument, 115–17; writing assignments, 36–37, 81–85; writing as tool, 45–46
style, in scribal writing, 60
subject, knowledge and expertise, 64–65
successes, in collaboration, 178–82, 185
supra-textual graphical elements, 122
syllabi, 36–37
syntax, 63, 68; and subject-matter knowledge/expertise, 59, 64–65

tables, 118, 124
teaching, 14, 105, 179(fig.); critical thinking, 75–76; department objectives, 76–77; disciplinary writing, 13–14; emphasis in, 65–66; gender dynamics of, 136–37; instructional activities, 66–68; interdisciplinary team, 23–24, 91–92, 93–99; and understanding, 55–56; writing, 45, 72–73
teaching style, gender and, 134
team reading, 68
team teaching: interdisciplinary, 91–99; socialization, 151–52
teamwork, 5, 24, 110
technical communication textbooks, on graphical elements, 118, 121–22, 124–25
Technical Communication courses, 61–62, 133; instructional emphasis, 65–66
technical sciences, gender awareness, 11–12
technical sketching, communication in, 112–13
tensions, in interdisciplinary collaboration, 9–10, 12, 24–25
theory, in engineering writing, 47
thinking, technical, 107
third culture, 12, 19, 35(n1); creating, 30–34
time-related issues, 29, 87
tools, shared, 88
transactional writing, 79, 80
Tribal Critical Theory, 165
trust, 12, 176, 185–86; of students, 142–43
truth, science and, 89
turnover, consultant, 182

universality, 187; assumptions of, 20–22
utility, of communication, 141–42

values, 21, 67, 151, 158; and hierarchy, 164, 165
visual: in engineering discourse, 121–22; in mechanical engineering, 107, 113; and written texts, 131–32
visual communication, 118–19
visualization, 22, 120; data, 118, 128
Vygotskian Perspectives/Activity theory, 154

WAC. *See* Writing across the Curriculum
WAC/WID work, 3, 10
ways of knowing, 23
WID. *See* Writing in the Disciplines
working relationships, 138
workplace, 4–5; in classroom settings, 41, 43, 44, 47
workshops, communication consultants, 152
worldviews, 169, 170; disciplinary, 87, 176; knowledge and, 157–58
writing, 3, 18, 23, 37, 38, 48, 50, 109; in classroom, 46–47; in chemical engineering class, 40–42; in civil engineering, 51–55; critical thinking skills in, 74–76; cross-cultural, 62–69; in design process, 114–15; electrical engineering class, 43–45; evaluating, 61–62; in exams, 97–98; instruction in 55–56, 69–70; laboratory settings, 94–95; learning and, 39–40, 42–43; in mechanical engineering, 105–6; rhetorical concepts of, 56–58; rhetorical vs. scribal, 10–11, 36; scribal concepts, 58–62; teaching, 72–73; as tool, 45–46; and visual, 131–32
Writing across the Curriculum (WAC), 3, 16, 105
writing assignments, 25, 36–37, 41, 43, 73, 146; in electrical engineering, 44–45; expressive and transactional writing, 79–81; laboratory memo, 100–104; in mechanical engineering, 111–15, 118–19
writing-to-communicate (transactional writing), assignments in, 79–80, 81, 83–85
writing-to-learn (expressive writing), assignments in, 78–79, 81–83
Writing Emphasis/Writing Intensive Programs, 38–39
writing fellow programs, 6–7
Writing in the Disciplines (WID), 3, 29, 30, 71, 85–86, 134; curriculum structure, 73–74
written, vs. visual, 107

zones of proximal development, 4–5

www.ingramcontent.com/pod-product-compliance
Lightning Source LLC
Chambersburg PA
CBHW060521080526
44586CB00012B/561